Construction UK:
Introduction to the Industry

Ralph Morton

Blackwell
Science

© 2002 by Blackwell Science Ltd,
a Blackwell Publishing Company
Editorial Offices:
Osney Mead, Oxford OX2 0EL, UK
    Tel: +44 (0)1865 206206
Blackwell Science, Inc., 350 Main Street,
Malden, MA 02148-5018, USA
    Tel: +1 781 388 8250
Iowa State Press, a Blackwell Publishing
Company, 2121 State Avenue, Ames, Iowa
50014-8300, USA
    Tel: +1 515 292 0140
Blackwell Science Asia Pty, 54 University
Street, Carlton, Victoria 3053, Australia
    Tel: +61 (0)3 9347 0300
Blackwell Wissenschafts Verlag,
Kurfürstendamm 57, 10707 Berlin, Germany
    Tel: +49 (0)30 32 79 060

First published 2002 by Blackwell Science Ltd

Library of Congress
Cataloging-in-Publication Data
is available

ISBN 0-632-058528

A catalogue record for this title is available
from the British Library

Set in 10/12.5pt Palatino
by DP Photosetting
Printed and bound in Great Britain by
MPG Books Ltd, Bodmin, Cornwall

For further information on
Blackwell Science, visit our website:
www.blackwell-science.com

Contents

Acknowledgements

The author and publishers would like to thank the following for kind permission to use the materials described:

W. W. Norton & Company for the drawing of the Pantheon in Rome on page 104, and for the quotation on page 102 from *Why Buildings Stand up: the Strength of Architecture* by Mario Salvadori.

The Construction Industry Training Board for Figure 4.5 and Tables 5.1, 5.2, 5.3 and 5.4.

The Centre for Alternative Technology for Figure 9.1.

MACE Ltd for information used in Figure 8.3.

Building published by Building Publications for Figures 3.3, 5.2, 5.3 and 5.4 and Table 9.1.

The Controller of Her Majesty's Stationery Office for crown copyright material taken from: the Construction Statistics Annual (Figures 2.1, 2.2, 2.3, 2.7, 3.3 and Tables 2.1, 2.2, 2.3, 3.1, and 3.2); and from the Housing Construction Statistics Annual (Figures 2.4, 2.5, 3.1, 3.2, 4.1, 4.2, 4.3, and 8.2).

Greenpeace International for the photograph on page 194.

The Peabody Trust for the photograph on page 170.

Some of the material used in Chapter 5 appeared in a different form in a paper by the author given to the Twelfth Bartlett International Summer School in Moscow, 1990. Parts of Chapters 4 and 6 draw significantly on the work of the Japanese historian Akirah Satoh in his book *Building in Britain* (see bibliography).

The author would like to thank all those who assisted with advice and information, including George Guy, Regional Secretary of UCATT, for a very helpful discussion on employment in the industry.

Preface

There are approximately one and a half million people employed in the construction industry; there are over 160 000 firms, from giant corporations to one man businesses; between them they produce £65 billion worth of construction work each year. It is a vast, complex and important industry. It has created and continues to create, for good or ill, the physical environment in which most of us live.

To do it full justice in 200 pages is not possible. What this book tries to do is to describe some of its features in outline – its structure, its markets, its people, its modes of operation; but also to introduce readers to some of the debates and controversies concerning important issues which have been going on within and about the industry for very many years.

An understanding of the industry's particular characteristics requires at least some knowledge of their historical development. Any short historical account is bound to be superficial, but in the case of construction it is possible to identify some key periods of change and powerful trends which do help us to understand why some of its characteristics and some of its problems became so deeply embedded.

Many of the issues raised are controversial and the personal views expressed (which would need much more space for a fully argued justification) may of course be challenged; they should at least provide stimulation for further investigation and debate.

There is an enormous amount of information available on construction in official publications, in books, journals, and now, in burgeoning quantity, on the internet. The suggestions for further study at the end of most of the following chapters are intended only as suggestions for a way into all that – as a beginning to an exploration of an intriguing subject.

About the Author

After graduating from Oxford University in 1959 Ralph Morton worked for a while as an economist in industry before moving into teaching. He taught for several years in schools and further education both in the UK and the USA, and then took up a post as Lecturer in Economics at the former Liverpool Polytechnic. He joined the Department of Architecture as Principal Lecturer in 1974, teaching economics and social aspects of architecture on undergraduate and postgraduate courses in architecture, planning and housing. His main research field was the economics of housing; he was awarded an MA and PhD by the University of Liverpool and published many papers on housing economics, housing policy and on architectural education.

Ralph Morton was appointed Professor and Director of the School of the Built Environment at Liverpool John Moores University, from which he retired in 1990. He co-authored, with David Jaggar, *Design and the Economics of Building*, published by Spons in 1995 and produced an English edition of *Building in Britain* by the Japanese architectural historian Akira Satoh, published by Scolar Press also in 1995.

Triumphs, Troubles and Reports

- The building of Westminster Palace – the best and the worst of construction
- The rise of the modern construction industry
- The modern critique – Latham and Egan
- Persistent problems – a backward glance

The story of a great building

The New Palace of Westminster – the British Houses of Parliament – was built in the middle of the nineteenth century to the design of architect Charles Barry, by some of the country's major contractors. It used new engineering techniques and machinery as well as the skills of hundreds of different traditional craftsmen and a huge manual labour force. It was a great triumph of construction and soon became the representative symbol not just of parliament but of Britain, recognised throughout the world. The queen was delighted and Barry received a knighthood. It remains a powerfully symbolic building today even if the Gothic style gives a sense of past rather than future glories.

But there is another side to the story. The construction was expected to take six years and cost £700 000. It actually took nearly 30 years and cost around £2 million (equivalent to at least £500 million today). There were disputes of every sort, starting even before construction began. There were arguments over the initial design competition; arguments over the estimates; arguments over the architect's fees. There was a dispute over prices with the contractors; there were supply problems with the specified stone. David Reid, appointed to design the heating and ventilation system, fell out with the architect and the two became bitter enemies. One of the foremen swore at the masons, who promptly went on strike (for thirty weeks!).

The history of New Westminster Palace was certainly a history of triumph over difficulties, but it was also a catalogue of the sorts of problem that plagued the industry not just through the nineteenth century but through the twentieth as well. However in the end it was built, and it became loved and admired by its users and its thousands of visitors.

At the other end of the scale from the grandeur of Westminster, and not far away, housing of appalling quality was being thrown up; it was the same in every industrial city. Not only was much of it cramped and dingy, airless and cold, it was badly built. The term 'jerry builder' came into common use to describe the perpetrators of the worst abuses.

The nineteenth century construction industry, if it could be called an industry at all, was a complex and confusing conglomeration of businesses and individuals capable at all levels of producing work of the highest quality – but also of generating some of the worst urban environments the world had known.

There have been many changes since those days, mostly for the better. We do not see building anywhere near the worst the Victorians could produce and we see much that is well built and technically far more sophisticated. Yet many of the less successful characteristics of the industry so well illustrated by the building of Westminster seemed to have become a permanent part of the construction system.

That is why, in trying to understand today's construction industry, some understanding of its past is essential. In the 1990s two major reports, known generally as the Latham and Egan reports,[1] were produced which identified a number of problems in the industry; they made many recommendations, some of which have been effectively implemented.

Most of the criticisms and many of their recommendations had been made time and time again over the previous 100 years, which raises some interesting questions. Why, when so much has changed, has so much remained the same? Why do the same criticisms keep on surfacing, when the industry is so obviously advancing in many ways? Will things really change this time? In the rest of this book, after looking at the structure of the modern industry and its markets, we look at some of its characteristics in their historical context, hoping to answer some of those questions and to gain a deeper understanding not only of how the industry operates but why it operates in the way that it does.

First, though, in this introductory chapter, we look very briefly at the broad historical background and then, after summarising the Latham and Egan reports, show how many of the issues they raise go right back to the time of Charles Barry and the building of the Palace of Westminster – and in some cases even earlier.

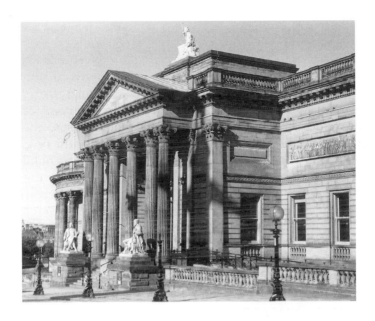

Photos (1) & (2)   Grandeur and squalor: Victorian towns generated both side by side. Two views of Victorian building in Liverpool. In the second picture each floor (including the cellars) housed at least one family, usually more.

Rise of an industry

Victorian construction had more in common with today's industry than it had with the methods of previous centuries. Somewhere around the beginning of the nineteenth century a radical change took place.

Building had for centuries been the work of independent craftsmen working directly for a client. Now a new type of firm developed, the general contractor, and with it a whole new method of organising the construction process. Joiners, painters and bricklayers now worked in their hundreds for big firms or as subcontractors to others. We look at some of the details and implications of this transformation in later chapters; the point to stress here is that there *was* a major change even though old practices lingered on and still do; the individual tradesman working on his own, for example, was to remain a permanent feature of the industry.

The new forms of construction developed in response to new forms of demand and they transformed the face of Britain. In 1801 no town in Britain, with the exception of London, had a population of more than 100 000 people. London, with about a million, was already one of the three biggest cities in the world. Liverpool, the second largest in England, had a population of 80 000 and Manchester 75 000.

Outside the capital every town and city in the country consisted of streets of small houses and shops, stables and workshops. There would be very few large buildings apart from the church or cathedral, prison, castle and maybe windmills and market halls. They would all have been built over quite long periods by teams of craftsmen. Beyond the towns, scattered across the countryside were the grand mansions of the rich and the farms and villages of the rural workers – all built by local people in local styles with local materials.

By 1850, London had grown to two and a half million, Liverpool to 376 000, Manchester to 303 000. There had been a vast increase in the area of housing (most of it quite dreadful); huge factories and warehouses dominated the cities. As the cities became wealthier, more and grander buildings were constructed – such as the magnificent town halls in Leeds, Bradford, Manchester and Liverpool; new law courts, bank buildings, multi-storey offices, concert halls, and theatres.

There had already been a vast expansion of civil engineering works – reviving skills hardly used since the days of the Roman empire; the network of canals with their locks and aqueducts, new docks in London, Bristol, Liverpool and Newcastle. Later, as the importance of hygiene was slowly recognised, came new water systems, sewage systems, better paving and drainage; and then in the second half of the nineteenth century the railways – with their miles of track, bridges, tunnels, viaducts, great cuttings and of course the stations which in the major cities were built on a grand scale.

The old ways of building could not have dealt with this massive demand; new forms of organisation were necessary for new forms of construction. For the first time large contracting and civil engineering firms were set up by enterprising people – often already in the building trade – who saw challenging opportunities in the growing scale of projects needed. The new industry was led by an extraordinary group of men with drive, energy and imagination whose achievements still impress us – even if the worst of Victorian building left a lasting legacy of degraded environments.

During the 150 years since those early days, the industry has evolved in many ways. Some businesses grew in size and their number increased. There were always many failures – the industry has always had one of the highest rates of bankruptcy – but some firms grew and prospered, developed their expertise and survived well into the twentieth century. The technology of construction changed but very slowly and not very dramatically. New materials came into use: concrete, steel and plastics; old materials were improved – particularly glass and iron – and were mass-produced in new ways. More components of buildings were prefabricated. Trade unions developed and 'masters and men' waged a long series of running battles over conditions, wages and forms of employment – usually leading to an uneasy truce.

Yet despite its achievements, throughout that long period the industry seems often to have been at the centre of controversy; there were complaints from government and clients of corruption, inefficiency, poor standards and excessive costs. Towards the end of the nineteenth century and throughout the twentieth century the government took increasing interest; more regulations were imposed, investigations undertaken, reports written; eventually there were even plans for nationalisation.

The Latham and Egan reports can be seen as the latest in that long succession of investigations, many of which as we see below identified similar problems and made similar recommendations. Sir Michael Latham and Sir John Egan set a series of challenges to the industry and gave us an 'official' view of what are now seen as some of its strengths and many of the weaknesses. The reports have been widely discussed, welcomed, criticised – and in some areas ignored. A brief review here of the critical issues which they raised may be a useful introduction to the extended discussion of the industry in later chapters of the book. When we compare some of the recommendations with those of earlier comments we can see perhaps how deep-seated some of the problems have been.

The Latham and Egan reports

In 1991, Sir Michael Latham was commissioned, in a joint venture by government and the industry, to conduct a review of the 'procurement and contractual arrangements in the UK construction Industry'. An interim report, *Trust and Money*, was published for consultation in December 1993, and the final report, *Constructing the Team*, in July 1994. The reports identified a wide range of weaknesses in current procedures. Most of these had already been separately recognised in the industry and indeed discussed in earlier official reports but Latham linked them together and set out an agenda for reform.

The report made over thirty specific recommendations, which can be summarised in a few categories:

● Government should take the lead in improving clients' knowledge and practice, particularly of how to brief designers and select procurement methods
● The whole design process should be reviewed and the link between design and construction improved
● Building contracts should be simpler, clearer, more standardised and less likely to lead to disputes
● There should be simpler faster means to resolve disputes where they do occur
● There should be a Construction Contracts Bill, outlawing some unfair practices, the introduction of adjudication as the normal method of dispute resolution and the establishment of trust funds for payment
● Government should maintain lists of approved consultants (designers, etc.) and contractors for public sector work
● The traditional methods of tendering should be revised and improved
● Training and research programmes should be rationalised and improved
● The industry should aim for a 30% reduction in costs by the year 2000.

Although many of the Latham recommendations were accepted and followed up, including the enactment of a Construction Act (discussed in Chapter 7), only a few years later, the Deputy Prime Minister, John Prescott, set up another committee – this time a 'task force' – under Sir John Egan of the British Airports Authority to advise

'from the client's perspective on the opportunities to improve the efficiency and quality of delivery of UK construction, to reinforce the impetus for change and make the industry more responsive to customer needs' (Egan Ch 1)

The Egan report, *Rethinking Construction* published in 1998, was shorter and sharper but more radical than Latham. The language was different, the criticism harsher and it implied a total change in the industry's culture. Many of the recommendations of the two reports were in effect very similar. But whereas Latham seemed to look for reform within the old traditions, Egan was proposing a revolution, or so it seemed.

The report identified what it called five key drivers for change:

- committed leadership
- a focus on the consumer
- integrated processes and teams
- a quality-driven agenda
- and commitment to people.

It set specific targets: an annual reduction of 10% a year in construction costs and time; a reduction in defects by 20% a year. The industry would need to change its methods radically, to create an integrated project process (we examine what this means in Chapter 8). It needed to improve working conditions dramatically and improve its management and supervisory skills.

The immediate impact of Egan was considerable. The report was widely discussed; new bodies were set up to push the ideas forward such as the Movement for Innovation (known as M4I), which was linked to the already existing 'Best Practice Programme'. The government moved towards forcing all public sector and publicly supported bodies such as its own departments, the health service and the housing associations to become 'Egan compliant'. The Auditor General's office produced its own report, *Modernising Construction*, showing how the Egan principles were to be applied throughout the public sector.[2]

What the long-term impact of all this activity will be remains to be seen but at least the industry, the government and the major clients can no longer be accused of not really trying. Nevertheless if change is to become permanent and spread across the industry, there needs to be an understanding of underlying factors that have made the very issues identified by Latham and Egan so difficult to resolve in the past. Most of these will be discussed in the following chapters but to show just how long some of the areas of concern have persisted, the next section looks at a few examples of their long history.

Persistent problems

The most commonly recurring complaints about the process of construction are that it costs too much, takes too long and produces poor quality. Of course this has never been true of most building and construction work. It is always the bad news that makes the headlines and the hundreds of buildings, big and small that are good value, good quality and built in time are barely noticed The very best are recognised in building and design awards but the rest are just built and occupied. The dozens of motorway bridges which function perfectly may be major feats of engineering – but the little footbridge that bounces is headline news for weeks.

The unfortunate thing is that often the projects that do have problems also have a high political profile. Recent examples are Portcullis House (the new offices for members of Parliament) and the British Library. Sadly, the Millennium Dome was too overwhelmed by other problems for many to notice that it was a highly successful and innovative piece of design and construction.

These overruns in cost and time as well as poor quality (not a problem with the Houses of Parliament) have been attributed to a host of other inefficiencies in the way buildings are procured, the ways they are built, the organisation of the industry and the training of the workforce, all problems identified by Latham and Egan, but all identified many times before. Just a few examples will serve to show how intractable some problems seem to have been.

Issues of procurement and contract

Both Latham and Egan criticised the methods of procurement used in the industry, that is the way contractors are selected and the relationship that is set up between client, designers, contractors and subcontractors. Both attacked particularly the way contractors are chosen – the system of tendering. Latham for example criticised long tender lists. Egan was more forthright:

'Cut-throat competition and inadequate profitability benefit no-one.'

Witnesses made exactly the same point to the Select Committee on Public Works *in 1813*. As we will see later, the battle was lost then. But over the next 180 years the arguments have been repeated over and over again.

In a letter to *The Builder* (the original name of *Building*), in 1847 a correspondent wrote:

'Sir – As your columns are at all times open to represent the evils and expose the present system of competition, injurious alike to all the parties concerned, I beg to forward in your valuable journal, the following almost unparalleled instance of the evils above referred to in the contract for the erection of eight houses for the Peckham Building Investment Company on some land adjoining the Licensed Victuallers Asylum, Old Kent Road. The plans and specifications were made by Mr George Smith and the tenders were delivered on the 28th of last month.'

This was followed by a list of 15 tenders submitted ranging from £1943 to £3920, that is, the highest was more than twice the price of the lowest. The writer may have thought it 'unparalleled' but it is a situation that has been repeated thousands of times since.

A thorough examination of the situation by a government committee in the 1940s and another in the 1960s both concluded that continuity of contracts was more important than going out to competitive tender for each job. The historian Marian Bowley wrote in 1966 after a detailed study of practice in the industry:

'It is high time that an end was put to sacrificing the possible improvements in efficiency ... to the idol of competitive tenders on finished designs'[3]

And yet nearly forty years later the same points are made   the same arguments go on.

The need for better training

Shortage of skilled workers and the inadequacy of training were a serious concern at the end of the first world war, and the problem continued, becoming even more severe during and after the second world war. A report in 1943 complained of the 'casual form of engagement which was formerly the most unsatisfactory characteristic of employment in the industry'; one of the consequences of this had been that although apprenticeship schemes existed, 'they followed no comprehensive and systematic plan'.

It therefore recommended 'the setting up of a training and apprenticeship council for the industry'.[4] That recommendation was followed then but the situation did not improve. The Construction Industry Training Board was set up in 1964. Yet 30 years later, there were complaints in the industry of skills shortages, Latham called the need for discussion of training reform 'a matter of urgency' and Egan said there were 'significant gaps' in training which had to be filled. The point here is not that no advances had been made over those years, they certainly had. The point is rather that the industry always seemed to its critics to be one step (or more) behind in developing programmes appropriate to current conditions.

The need for standardisation and prefabrication

Shortly after the First World War and the passing of the series of housing acts from 1919, the government became concerned at the high cost of 'working class housing'. The departmental committee of 1921 warned local authorities against introducing too many types of building because 'all classes of builder put a high price on a cottage of a type with which he is not familiar not because it really would cost more ... but because he has no experience and does not feel able to estimate readily the difference between this cottage and the one he understands'. The committee recommended that measures be taken to secure the adoption of new simple types of houses evolved to substitute for the less economical types in existing contracts. Local authorities were to be allowed to make adaptations to use local cheaper materials but the standardisation of windows doors and sanitary goods and internal fitments 'should be rigidly enforced'.[5]

A few years later a committee on New Methods of House Construction was set up and produced a series of reports, which among other things recommended greater use of concrete and steel construction systems; they wanted to see greater use of 'what may be called factory production of houses' but were concerned that few practical and economic systems had been developed. They seemed to hold out hope for the development by one Commander Burney who was trying to produce a new light material which could be pressed into moulds to make panels for walls; history doesn't seem to record what happened to Commander Burney or his material.[6]

In the middle of the Second World War, a mission sent to study systems in America urged the wholesale reorganisation of the British building industry; among many other specific points, it recommended:

- simplification of building design for greater standardisation and mechanisation of constructional work
- much more use of factory produced units and assemblies.

Fifty five years later, Egan recommended precisely the same change of direction.

Integration of the building team

This was a key concept in the Latham report, as its title shows, and it is perhaps *the* key idea in the Egan report. Yet in a report on systems building in 1964, A.W. Cleeve Barr had written:

> 'Industrialisation is a process of change which involves the substitution of hand labour by machines both on site and in the factory. It involves the use of new techniques and new materials and the use of traditional materials in new ways. And it involves – for its fullest exploitation – new ways of co-ordinating the building requirements of clients and new contractual and working relationships between clients, architects, builders and manufacturers'.[7]

This last sentence points more or less exactly to what Egan was to describe over thirty years later as 'partnering'.

These examples could be extended at length. But the point should now be clear – the same issues have been raised over and over again. Of course many things have moved on, much has improved. But in some areas it seems that little has really changed – except at what Sir John Egan calls the 'leading edge' of the industry.

So there is something of a puzzle and many questions without obvious answers. What is it that has prevented the industry as a whole from responding to these recurrent appeals for change? Has it in fact changed much more than it has been given credit for? Do the criticisms misunderstand the essence of the industry? Is there a problem of competence? Or are economic forces in the construction industry too strong to allow real change to take place?

In investigating and describing the many facets of the industry we will approach these questions in different contexts and present much of the material that will help to provide answers. But this book is not just an exploration of problems – it is a survey of the industry that, hopefully, will put these much-publicised problems into perspective. The next section outlines the scope and organisation of the rest of the book.

An overview of the book

The following chapters fall into four groups.

Chapters 2 and 3 cover the characteristics of the *demand for* and *supply of* construction works. Chapter 2 describes the organisations that require the work done and the type of construction needed, as well as some of the underlying economic forces which determine the levels of demand. Chapter 3 describes the structure of the industry that supplies that demand, the many different kinds of organisations that make up the industry and again the forces that are constantly leading to change.

Chapters 4 and 5 describe the *industry's workforce* – the people who actually make things happen – workers on site, managers, designers, engineers, and surveyors. It examines many of the major issues of training, relationships, working conditions and organisation. Both chapters include some historical background as an explanation of some of the peculiarities of industrial relations and what seems to many outside of the industry the peculiar set of relationships between the various groups of people working in their different specialisms.

Chapters 6, 7 and 8 look at the way the industry organises the *procurement and production process* – that is the way buildings and other types of construction are actually provided. Chapter 6 follows in outline the history of the system's development because, as in the case of industrial relations, the historical perspective really does help us to understand why the industry operates in the way that it does.

Chapters 9 and 10 set the industry in a *wider context*. Chapter 9 is concerned with environmental issues – the use of energy in construction activity and buildings, the environmental impact of its use of materials, and the physical impact of the industry on city and countryside. Chapter 10 brings together many of the early themes by summarising the relationship between the construction industry, the state and society as a whole. In every chapter of the book we will have come across the importance of the many links between government and construction – so this final chapter will act as a review of most of the topics dealt with along the way.

For further study

The Egan report is of course essential reading as is the Audit Commission's *Modernising Construction*. On the history of earlier reports see the new book edited by David Langford and Mike Murray: *Construction Reports 1944–98*. The building of the Palace of Westminster is described in P.W. Kingsford's *Building and Building Workers* (1973) and in a fascinating article in *Building's* 150th Anniversary Supplement (1843–1993) *The Saga of Westminster Palace*.

2 Who Needs Construction?

- The peculiarities of demand for construction
- The general picture: categories of demand
- The industry's clients, public, private, corporate and individual
- The volatility of demand – some causes and consequences

Introduction: some peculiarities of construction demand

The main purpose of this book is to describe and explain the characteristics of the modern construction industry. However, before looking at the industry itself, its organisation and its methods of operation, it seems logical to look at the nature of demand for the industry's products. There are particular characteristics of that demand – for houses, hospitals, office buildings, schools, roads, bridges and a multitude of other structures – which explain to a large extent why the industry is the way it is.

'Demand' in this context means the value or the quantity of construction projects that people or organisations want and are prepared to pay for. In the second section of the chapter we will look at estimates of this value, and how the totals can be broken down into different categories. The third section describes the characteristics of the major groups of the industry's customers (usually in the context of construction referred to as clients). The final section, before a brief summary, examines how demand fluctuates over time, for this, as will be seen later, is an extremely important influence on the way the industry has developed.

But first, it is worth pointing out that, at a general level, demand for construction has two characteristics that make it different from demand in most other modern industries. First, it is very often 'bespoke' demand and secondly, the level and type of demand cannot easily be controlled by the industry itself. Each of these is discussed below.

'Bespoke' versus 'off the peg'

There was a time when anyone wanting a new suit or dress or even a shirt would naturally go to a tailor and ask him to make it. Every town had dozens of 'bespoke tailors' who would provide this service. This is still done of course but most us are more likely to buy clothes ready made 'off the peg'. The majority of consumer products are now like this – ready made in thousands or millions. Construction however is still very much a bespoke industry. Clients define their requirements – in terms of a type of building say – and then have it built to their specification. The industry has developed its organisation to meet demand arising in this way.

There are many exceptions. The most obvious one is housing where housebuilders provide the finished (or nearly finished) article – and the consumer chooses which type he or she wants and can afford. And there are many other forms of speculative building – such as office blocks, general retail space and factory units – which are built first and then sold (or let); but apart from these important exceptions, orders for construction are for specific projects of specific types in specific places.

An interesting question, which is raised by the Egan report and will be looked at in Chapter 8, is whether construction should and will move in the same direction of other industries – like clothing – towards much more 'off the peg' production.

Manipulating demand

In many other industries, a substantial proportion of effective demand for their products is created and managed by the producers themselves, through extensive marketing and the application of new technologies. This is true, for example, of cars, television, washing machines, CD players, video cameras, package holidays and a thousand other products. In rich countries at least, even the demand for basic goods such as food, is partly created and manipulated by the producers themselves. So-called luxury and junk foods are eaten because they are available and vigorously marketed, not because they are needed (we spend in the UK some £1.2 billion a year on ice cream!). Of course overall levels of demand still depend ultimately, in the ways economic text books describe, on factors such as the size of the consumer population, incomes, income distribution and relative prices – but the power of marketing today is enough to determine the constituents of that overall demand.

Demand for buildings again is in general rather different. It is a secondary or derived demand, determined by the demand and output of other goods and services. The number of factories required is the result of decisions by manufacturers about the goods the factories will produce – which in turn depends on the demand for those products. The number and type of hospitals and school depends on decisions by government, themselves based on need and available finance. Office development depends on the real or perceived demand for office space – which depends on the amount of work available for the companies using them. One of the consequences of all this is that although the industry can certainly market its services, it cannot easily *create* demand for its products; there is no equivalent to the latest version of a computer game or a new model of car.

Again a possible exception is the demand for housing, which, as pointed out above, is much more like a consumer product. But even the house-builders are less able to manipulate total demand than most other manu-facturers; the demand is determined by population structure and movements, by incomes, household formation and mortgage conditions. The shift of population from some regions of the UK towards the South and the change in household structure towards more single people living alone are not factors under the control of the industry. The housebuilders can build attractive new houses in certain localities and entice people to them – but they cannot defeat the underlying flow. Housing associations which have built new houses in some areas, particularly the Northeast, have discovered this to their cost; in one area new houses were demolished a few years after they were built, as there was no demand for them.

Putting both characteristics together – the fact that demand is largely 'bespoke' and that it is derived from demand for other products, it follows that the industry as a whole has to respond to the character and pattern of demand, *as it arises*; it is much less able to manipulate demand for its pro-ducts than are many other producers. It has therefore had to develop par-ticular structures and methods of operation that can respond appropriately to clients' requirements.

The next section looks at the scale of those requirements – the total demand for construction – and at the relative significance of the different categories into which it is usually divided.

Demand for construction: the general picture

Some definitions

The most comprehensive and accessible source of information about the structure, levels and changes in demand is contained in the series of con-struction statistics produced by the Government Statistical Office and since 2001 published as the *Construction Statistics Annual*. This chapter relies heavily on those statistics as well as using some information from other sources.[1]

As is always the case with this sort of information, it is possible to classify it in many different ways. The official statistics are analysed in two main categories – a broad classification by *type of client* – public or private – and a more detailed classification by *type of construction* (roads, schools, hospitals, etc.). For each of these categories, there are two sets of figures which might be used as measures of demand in the sense defined above: first, *new orders* obtained by contractors and second, actual *output*; obviously the first of these is more strictly a measure of 'demand' at a point in time but each set of statistics also contains different sorts of information. The new orders statistics for example do not include figures for repair and maintenance work, while they do give an interesting breakdown in terms of size of project, so both sets are useful for different purposes and both will be used here.

Although on the face of it the division between public and private client seems fairly straightforward, there are some problems of definition because the boundaries between the two sectors are changing and becoming increasingly blurred. The public sector traditionally included:

● Government departments and central government agencies (such as the Highways Agency)
● Public utilities – such as the former gas, water and electricity boards
● The nationalised industries
● The Post Office
● Universities and other education institutions
● Local authorities and new town corporations
● Housing associations.

The private sector was virtually everything else – from the large industrial and commercial companies needing factories, offices, warehouses and other facilities down to the individual householders needing minor repairs to their homes.

There have been continuing shifts mainly from public to private and policies which have blurred the line between the two; some of the most important are:

● The private sector now includes the major utilities which have been privatised
● Most other former nationalised industries are now private companies
● The former polytechnics which became universities in 1989, once the responsibility of local authorities, are now independent private sector organisations
● Housing associations, formerly financed through public loans and public grants, were considered part of the public sector; now they are

financed partly through commercial loans and considered as private businesses. To make matters really confusing, they are, according to the definitions in the official statistics, still part of the public sector for the purposes of the new orders and output statistics but they are now included as part of the private sector in the housing statistics. They are discussed in more detail below as public sector organisations

● Finally although the public sector is supposed to include all work done for public bodies, work done under the government's Private Finance Initiative (again discussed below in more detail) is described officially as 'in principle' private sector work.

All this makes some of the published statistics difficult to interpret, particularly when looked at over a long period, as the meaning of public and private sector has changed. However, in spite of these difficulties, it is worth looking at the figures in the two categories as they indicate the continuing significance of the government as the country's largest single client for construction. As we will see throughout the book this fact, together with its control of economic policy and its powers of regulation, give the state tremendous influence over the industry.

The structure of demand – a summary

Tables 2.1 and 2.2 set out, in summary form, the basic elements of construction demand for two years a decade apart – 1989 and 1999. Table 2.1 shows the value of *new orders* for construction and Table 2.2 the value of construction *output*, which includes figures for repairs and maintenance. No allowance is made for inflation but it is the relationships between the categories of expenditure in each of the two years that concern us here – so that is not too much of a disadvantage. These tables show how the composition of demand changed in that decade and also enable us to identify some important characteristics of demand that have a significant impact on the way the industry works.

Public and private sector

Looking first at the totals in Table 2.1, it can be seen that the total public sector spending is a significant part of the whole, but that it reduced considerably between the two years. In 1989 the public sector was responsible for just over a quarter of new orders (£7077 m out of the total, public and private, of £27 145 m). In 1999 the share was not much over a fifth. However this is still a very high proportion for what is in effect one major client – the state.

Table 2.1   Construction orders from public and private sectors 1989 and 1999

	Public sector £ million		Private sector £ million	
	1989	1999	1989	1999
New housing	872	969	6497	5901
Infrastructure	2369	1495	592	2700
Industrial	351	101	3049	2558
Education	610	1136	179	393
Health	824	635	295	411
Offices	519	390	5271	3566
Entertainment	382	435	1418	2224
Shops & garages	101	65	2491	2167
Other	1049	512	276	421
Total	7077	5738	20068	20341

Table 2.2   Construction output by major categories 1989 and 1999

	Public sector £ million		Private sector £ million	
	1989	1999	1989	1999
New housing	966	999	7117	6163
Infrastructure	2716	2270	1813	3893
Other new	3903	4509	13317	15833
Repairs & maintenance:				
Housing	4943	6459	8149	9825
Other	4982	5364	4755	8334

Housing

New orders for public sector housing (Table 2.1, line 1), which in the past have been as high as orders for the private sector, are now a small proportion of total housing built though there was a slight increase between 1989 and 1999 – from 12% to 14%.

Infrastructure

Infrastructure new orders (Table 2.1, line 2), that is for work on roads, rail, gas, electricity, etc. (see Fig. 2.1 for a breakdown of the figures in 1999) show the most dramatic shift, as a result of privatisation of the nationalised industries and utilities. In 1989 the public sector was responsible for 80% of infrastructure spending; by 1999 it was down to 35%. However infrastructure is still the largest category of public expenditure – now mainly accounted for by the road programme.

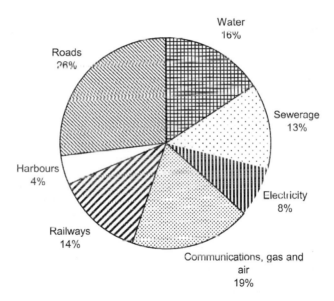

Fig. 2.1 Infrastructure spending 1999.

Other new construction

There is a very wide variety of buildings required by both public and private sector clients. The tables identify only a few broad categories. In the private sector as one would expect, the major categories are industrial buildings, offices and shops. The usefulness of the public sector figures is rather limited by the large 'other' category – though it does indicate the range of different types of demand.

Maintenance and repair

Turning now to Table 2.2, a summary of figures for *output* (as opposed to *new orders*), it is possible to see immediately the importance of repairs and maintenance. The pie charts (Figs 2.2 and 2.3 derived from Table 2.2) show the picture clearly. Maintenance and repairs represent approximately half of the total value of construction output. In the public sector the share is 60% and in the private sector it is about 40%. Looking back at the table itself, you can see that repairs to housing represent half of all the repairs in both sectors – and that the proportion has not changed much over the ten years. All this is highly significant for the industry – because most (but not all) repair work is relatively small scale, a fact which we will see in the next chapter. This accounts largely for the way the industry is structured – and also accounts for some of its problems.

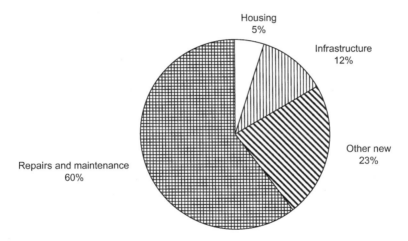

Fig. 2.2   Output for the public sector 1999.

The small scale of most construction work

Table 2.3 shows that much of new build work is also on a very small scale. It shows the proportions of selected construction categories in different value ranges. Only projects costing more than £25 000 are included as there are no accurate figures for the many thousands of very small jobs. Over 50% of public sector new work orders (excluding housing and infrastructure) are for projects worth below £100 000 as is about 50% of industrial and com-

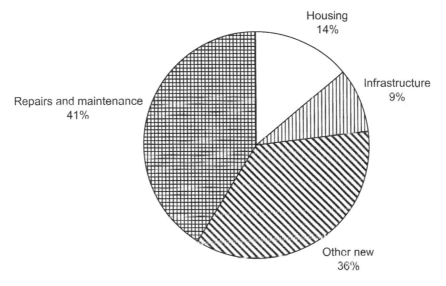

Fig. 2.3   Output for the private sector 1999.

mercial work. When one considers that £100 000 represents, say, only one small house, these figures are actually quite surprising.

This very general, statistical, picture of demand for construction has brought out several of its important characteristics; the next section describes the different client groups and some of the factors which determine their expenditure on construction – factors which lie behind the bare figures in the tables and charts above.

Table 2.3   New orders by value range and construction category

Value range £000	Public sector %		Private sector %		All infrastructure %
	Housing	Other	Industrial	Commercial	
Under 100	12.2	51.3	49.0	51.9	47.3
100–199	14.6	17.0	15.4	16.6	16.5
200–499	29.9	16.0	15.4	14.0	17.3
500–1999	34.9	12.6	15.1	12.1	13.6
2000 & over	8.4	3.1	5.1	5.4	5.3
	100.0	100.0	100.0	100.0	100.0

The industry's clients and the determinants of demand

This third part of the chapter is divided into three main sections: first an examination of the public sector and how its demand for construction is related to economic and other policies; secondly a look at some categories of private sector client and how they too are affected by policy; and finally an explanation of the Private Finance Initiative – which is to some extent dissolving the distinction between 'public' and 'private'.

The public sector

Central government

It can be inferred from Table 2.1 above that central government departments are directly responsible for much of the public sector work. Most of the public sector infrastructure spending for example is now expenditure on major road schemes. The former Department for Environment, Transport and the Regions (DETR) let over £500 million of road maintenance contracts during 1988 and 1999 as well as a number of major new road schemes. The DETR was probably the biggest spender on construction of all government departments, responsible not only for roads and some rail expenditure, but also for allocation of funding under all the government's special regeneration initiatives (such as the Single Regeneration Budget) and the large sums which have been received from European regional development funds in areas such as Merseyside and Northern Ireland.

The Department of Health is another big spender; in 2000 a large hospital building programme was well under way. As well as projected PFI schemes of over £4.5 billion, there was another £534 million-worth of schemes under construction; the largest hospital projects range between one and several hundred million pounds – so these are some of the largest and most complex schemes the industry is engaged in.

There is a wide range of other capital projects directly under central government control – the prison-building programme (Home Office), a number of new offices for the Inland Revenue. The Ministry of Defence has a total building programme of well over £1 billion in conventional contracts as well as many millions of pounds worth in projected PFI schemes. Then there are the major one-off schemes such as the new British Library, Portcullis House and the Millennium Dome.

The significance of Government as construction client is clear from the figures. But what determines how much is spent? This is a big question with many complicated answers. It is not too misleading however to suggest that there are two dominating factors leading to decisions to spend. One is the perceived requirement of a particular service (the need for more prisons, new offices for the tax inspectorate, new hospitals) together with the government's view as to how much can be afforded. The other is the government's perception of the impact of its spending programme on the economy as a whole.

In terms of the requirements of the different sectors of the public services, the amount of spending and its allocation will depend on a particular government's political priority. The Labour government has made it clear that education, health and transport are its top priorities. So at the time of the 2001 election there were plans to increase education capital spending from £2.2 billion to £3.8 billion by 2003/4 and health spending from £1.5 billion in 1998/9 to just below £3.5 billion (and 100 new hospitals by 2001). Public infrastructure expenditure is expected to rise to £6 billion, which will be £12 billion if PFI schemes are included – doubling the pre-2000 level. In fact so great will the impact of this spending programme be that many in the industry have questioned whether it can possibly be delivered.

The second major determining factor – certainly in the past though perhaps no longer so significant – has been the state of the economy as a whole. Under conventional 'Keynesian' economic policies, government spending and in particular capital spending were used to control fluctuations in the economy as a whole. So, in times of recession, as output and employment fell, governments would increase spending to stimulate economic activity. In times of boom when inflationary conditions threatened to get out of hand, governments reduced spending. For both economic and administrative reasons, the reductions and increases were often concentrated on construction projects. Sometimes these changes took place quite quickly and the construction industry frequently complained that in effect it was being used by government as a regulator of the whole economy – at one time flooded with orders at another finding itself short of work.

Recently the government has tended to rely on the Bank of England and its regulation of interest rates to control general economic fluctuations. However, this again has a particularly significant impact on construction. Changes in mortgage rates, as mentioned above, are one of the most significant determinants of demand for houses and general decisions to invest in factories, offices and other new building will be markedly influenced by interest rate changes.

We look at some consequences of this in the next section.

Local government

Local authorities are responsible for local roads and other transport facilities, schools, colleges and community buildings such as sports centres.
Decisions about what to build, where and when are taken at the local level
but the authorities' capacity to spend is virtually determined by the capital
funding and borrowing powers devolved to them from central government.
This is a situation very different from the United States, for example, where
a state or district can ask local electors to approve the issue of bonds to
finance specific projects.

There have been some shifts in the direction of independence in the UK
such as the possibility to tap into special government, European and other
funds (such as the Single Regeneration Budget, the National Lottery, and
the European Regional Development Fund. Access to these requires the
preparation of detailed bids, which involve much work and may well fail;
the final decisions again therefore are taken out of local government hands.

The Private Finance Initiative, discussed in more detail below may also
prove a new route for local government independence in capital expenditure. Currently there are many projects being planned and built under PFI,
including fire station and police facilities, schools and sporting facilities. All
these schemes however still require central government approval. The
availability of lottery funding has provided a further source of capital and is
already producing many new projects either for local authorities themselves or for partnerships between local authorities and local community
groups.

Total local government spending on construction has ranged from £4.4
billion to almost £5 billion per annum during the 1990s. Nearly a quarter
was spent on transport facilities, a third on education and a fifth on housing
– mainly refurbishment and conversion – and the rest on other facilities
including sport and recreation.

Not many years ago the local authorities' main capital spending was on
new housing but since then there has been a steep decline in housing
expenditure as the task of producing new low cost 'social' housing has been
increasingly taken over by the housing associations. Figure 2.4 shows how
important this change has been.

Housing associations

Some housing associations have origins back in the nineteenth century
when they were mainly charities, set up sometimes by wealthy individuals
(such as the banker George Peabody – founder of the Peabody Trust in

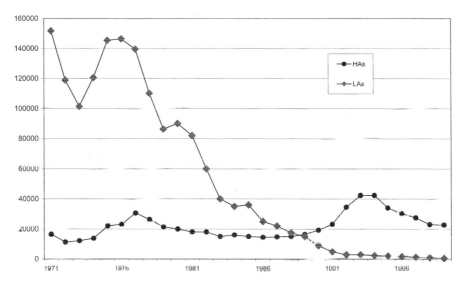

Fig. 2.4 Housing completions by housing associations and local authorities 1972–1999.

1856). But their modern period of rapid growth began only in the late 1960s and 1970s when a series of housing acts made them eligible to receive government subsidy to build low rent housing. In return for the receipt of subsidy, they were subject to control by government through the Housing Corporation. During the 1990s, the available levels of grant for house-building were reduced and the associations expected to raise much of their capital funds from banks and other financial institutions. They are now formally independent businesses (and with other similar organisations officially referred to as 'registered social landlords'). Many are still, however, also registered charities and not allowed to make profits. They are, as mentioned above, treated as private sector in the housing statistics – but they are still very much subject to financial regulation and other forms of control by the Corporation.

Of the 2000 associations, most are small; over half have fewer than 500 units (that is, flats or houses) under management and many of these have fewer than ten, operating in a very small area. But there are over seventy associations with more than 5000 units each and the biggest of these have tens of thousands. North British Housing Association, for example, which in spite of its name operates throughout the country has more than 44 600 units in management and develops about 4000 per year. Riverside is reaching a comparable size. Recently under the large-scale Voluntary Transfer Programme thousands of properties have been transferred from local authorities to associations or to newly established societies and will all require substantial renovation. As a whole therefore the housing associations and similar bodies are substantial clients of the construction industry, both for new building and for maintenance work.

They are however a special sort of client. Standards are high and strictly monitored; the associations themselves have acquired considerable expertise in dealing with contractors. They tend therefore to be demanding clients to work for but on the other hand, with very few exceptions, good payers and – so far – unlikely to get into financial difficulties. Many contractors, including former speculative housebuilders, have specialised in working in the sector and, in most parts of the country, development by even the smallest associations has proved an important source of work for local builders.

The housing associations have also become one of the vehicles through which the government has recently promoted new initiatives in construction procurement such as 'partnering'; these will be dealt with in later chapters.

The Private Finance Initiative

The Private Finance Initiative (PFI) is essentially one of a new range of public sector procurement methods (which are dealt with in detail in Chapter 7) but its impact has been to alter the whole character of the public sector as client and to force the industry to approach public projects in a new way altogether, leading as we see in Chapter 3 to actual structural changes in the industry – restructuring of firms, development of consortia and mergers. So it is a significant demand factor today.

The PFI was introduced under the Conservative government in 1992 but continued and was expanded and considerably refined by the Labour government after 1997. The essential idea is to transfer the funding and management of major public sector projects to the private sector. The new initiative was needed it was argued, 'because the electorate wanted a considerable and rapid improvement in the quality and availability of its public services but was not prepared to pay the extra tax'. The Chancellor of the Exchequer at the time, Kenneth Clarke, said, 'the private sector will take forward projects which the public sector used to undertake and do so more efficiently'. Private companies can draw on better design and management enabling projects to be completed more quickly and to a better specification;

in many cases they can go on to manage the project's operation as owners or holders of a management concession. Whether PFI schemes will in the long run turn out to be value for taxpayers' money is still contested by many and the whole system is bound to remain highly controversial, but the fact remains it is now an increasingly significant part of what used to be public sector work.

PFI is now considered as just one variant of the more general public–private partnerships with some versions referred to as DBFO (Design, Build, Finance and Operate) schemes. Basically the public sector body (for example area health trust or government department) invites bids to develop the required facility (the new hospital, prison or road scheme) and then to operate it for a fixed period of years (up to 60). The successful bidder – which may be a contractor or more likely a consortium of contractors, consultants, facilities mangers and others – contracts to provide the facility in return for an agreed annual fee (with the possibility of variations under agreed circumstances). The actual finance for the construction will be provided by the banks and other financial institutions though a new government-backed scheme is being developed.

The first projects under the schemes were roads, hospitals and prisons; it was a slow start; there were many teething problems as both contractors and the public sector bodies tried to sort out the complexities of the new contractual situation. There was much argument between the industry and the government over the way the system was developing. However in general the construction industry was, and remains, very enthusiastic, essentially reorganising itself in ways we examine in Chapters 3 and 7 to respond to the new way public sector clients now operate.

At the beginning of 2001 there were over 350 PFI projects under consideration, under negotiation or actually in progress. Approximately 70 of these were in the education sector including many new schools and colleges as well as refurbishment and expansion projects. The largest number of projects (over 150) was for the Department of Health and included several complete new hospitals as well as many smaller schemes involving redevelopment or construction of special units; the value of projects ranged from £630 million (at Bart's for the London NHS Trust) down to just over £1 million, most of them being at the lower end of the range. But in terms of contract value, the transport projects outweigh all the others, with the Channel Tunnel Rail Link and the Jubilee Line extension together coming to more than £7 billion – more than all the education and health projects combined. Some of the PFI projects are not in fact construction – they include for example installation and maintenance of computer and software systems – but most include both the construction and long-term servicing of buildings, roads or railways; there is no doubt about the importance of the initiative for the construction industry as a whole.

The private sector clients

There are so many different types of private sector client that it is possible to describe them only in broad groups. Four are considered below: the clients for small buildings, the major clients developing for their own occupation, the property developers and finally the private house buyers.

The clients for small buildings

As was shown above most projects are relatively small scale; it stands to reason that most of these – though obviously not all – will be built for small organisations or for individuals who require new premises or extensions and alterations to older buildings. This work provides the core of activity for the small to medium sized building firms generally operating locally. There is a great variety of potential clients – small manufacturing firms, local retailers, professional practices such as lawyers, accountants, doctors, vets and non-commercial organisations such as community groups and charities. But despite their variety and the variety of their requirements, they tend to have one characteristic in common – *they have no expertise in getting buildings built.*

For many, having building work done and particularly acquiring a whole new building may be the only construction project with which they have ever been involved. At worst this lays them open to unscrupulous practices but even at best they need some guidance through the minefields of gaining planning permission, selecting a designer, negotiating a contract and handling all the problems which every building project seems to generate. There have been many publications offering advice – such as the National Development Office's *Before you build: what a client needs to know about the construction industry* – published forty years ago. Advice is available from the Royal Institute of British Architects (RIBA), the Royal Institution of Chartered Surveyors (RICS) and other sources. But this lower end of the market tends to be ignored in discussions on the industry despite its importance in terms of the number of schemes and the value of work involved. It is a sad indication of the current situation that most debate about this sector has been in the context of concern about so-called cowboy builders (a debate which we will look at briefly in Chapter 10).

Major clients acquiring buildings for their own use

At the other end of the scale, there are large and powerful clients; it seems to be felt in the industry that the influence of these clients together with the scale of the projects they undertake has grown considerably over the last few decades, forcing even large contractors into a relatively subservient position.

One group consists of those organisations that have a continuing large-scale programme of construction for their own requirements. A major example is the British Airports Authority, sometimes referred to as a 'construction superclient' which spends around £450 million a year on construction work. Its individual projects are of enormous scale and complexity, currently including the £200 million extension at Stansted and the £2 billion Terminal 5 development at Heathrow. The company was a pioneer in establishing five-year (initially) 'framework agreements' with contractors and consultants who could then expect to receive a share of the work; contractors had to go through a rigorous selection process – and in fact some were removed from the list even after they had been initially successful. Well-known contractors were shocked to be removed and more recently BAA has introduced a system of annual tests to see if they still meet BAA's requirements. In February 2001 it announced that it would itself run the Terminal 5 project at Heathrow – thus becoming in effect both client and main contractor, with another 31 main subcontractors (or 'first tier suppliers') including some major contracting firms such as Amec working under them.

Other examples of large organisations with continuing long-term construction programmes are British Telecom, the supermarket chains and other major retailers such as Boots, the John Lewis Partnership and the newly privatised utilities such as the water companies. The supermarket chains alone are responsible for 20% of total spending on construction, which makes them collectively almost as powerful a client as the government. In 1998, the Office of Fair Trading investigated their treatment of construction industry suppliers 'as part of a wide-ranging look at whether they are abusing their buying power'. In 1996–1997, Tesco was the biggest single client with orders of £200 million while in 1998 the top place was taken by Sainsbury (over £300 million) with Asda not far behind. All had further expansion planned – Tesco intending to build a further 12 000 square metres of shopping area each year until at least 2003.

The hotel chains make up another group of clients with large-scale more or less continuous construction operations. Hilton, Granada (which owns

Travelodge), Whitbread and Balmoral International have all been reported as engaged on development programmes over the next few years.

The property developers

A second category of major clients is the group of property developers who have been generally responsible for the development and redevelopment of much of our city centres over the past fifty years or so. Some of their projects are built speculatively, that is without a specific client, in the hope that they will eventually be let – either to single or several clients at rents which will yield a good profit to the developer. Many of these projects and their developers became well known and sometimes notorious during the period of the property boom in the 1960s. One such was Centrepoint in London built by a developer Harry Hyams and kept empty for years after it was built. When it was eventually let it yielded Hyams an enormous capital profit. More recently One Canada Square at Canary Wharf was seen by many at the time as a potential white elephant – but with a new boom in demand for office space it has not only been fully let but become a prestige location.

There are many developers operating on a fundamentally speculative basis throughout the country. Grosvenor Estates, for example, has built office blocks in London (such as Old Broad Street) and shopping centres in Basingstoke and Dublin and was reported in 2000 to be contemplating a gigantic £500 million mixed development scheme in central Liverpool. Grosvenor, with British Land, Rosehaugh Stanhope, MEPC and many others are now seen as 'traditional' British developers. Recently (i.e. during the last ten years), a new specialist group of developers has appeared on the scene. One is an American company HQ Global Workplaces and another, a younger British company, Regus; these develop fully serviced offices, complete with furniture, IT equipment and even secretarial services, which companies can rent at very short notice and sometimes for relatively short periods.

At Canary Wharf, One Canada Square has now been joined by two new towers (together, at 44 storeys, the second tallest buildings in London). These towers are examples of the other type of developer's project – those built for specific clients. In this case, one of the towers is for the HSBC Bank and the other for Citigroup as its European headquarters. The developer in both cases was the Canary Wharf Group (and the contractor Canary Wharf Contractors!).

Many if not most of these developers, whether working speculatively or for particular clients, have operated closely with the same group of main contractors. At one time or another, most of the major contractors have themselves become developers when they have seen the possibilities of greater profit in that field and developers (as in the case of the Canary Wharf Group and now BAA) have become their own contractors. But more recently, many seemed to have retreated to their original core business as either contractor or developer (a trend discussed in the next chapter). So the picture is rather confusing; the point in terms of the focus of this chapter is that these large developments represent a particularly important form of demand for construction and have a considerable effect on the structure and operation of the industry.

The house buyers

Private house buyers make up the largest group, numerically, of the industry's customers. As has already been suggested, the housing market is very different from the others discussed above and requires a different sort of construction organisation. Housebuilders are themselves the developers, purchasing land, designing and building, and then marketing, the final product.

The analysis of demand for new housing in the private sector is now itself a major preoccupation, because of its financial and economic importance, for academics, mortgage lenders, city analysts and the housebuilders themselves. The main determining factors of variations in demand were mentioned briefly above. Some of these, such as birth rates and household formation, change only slowly; others such as migration from one region to another, household incomes and rising expectations may have effects in a shorter time – but still are changes that are measured over years. The one factor that can change the market within weeks is the mortgage rate, which is determined by the general level of interest rates controlled by the Bank of England. Although the Bank is now formally independent of government, it is still the case that the housebuilding industry is to a degree at the mercy of central policy decisions.

We have looked in the last two sections at what might be called the structure of demand – the way the total is divided among the different client groups – and we have looked at some characteristics of those groups. However, one of the most significant aspects of demand from the industry's point of view has always been its variability over time. The next section examines some of the facts and the issues involved.

Fluctuations in demand over time

It has often been claimed that construction has suffered more than any other industry from demand fluctuations. That would be a difficult point to prove but there is no doubt that alternating slumps and booms in construction have caused real difficulties, particularly when change has taken place quite quickly. This final section describes some of the fluctuations that occurred over the last decade of the twentieth century but the earlier decades have shown similar and in some periods more severe patterns of slump and boom.

Figure 2.5 charts total output over the period 1980–1999. The picture does not look particularly dramatic there, though it does show a clear cycle of slump and relative boom. Figure 2.6 relates the annual percentage change in construction output with the annual percentage change in gross domestic product (the standard measure of the country's total output of goods and services). This shows quite clearly (in fact probably exaggerates) the fact that fluctuations in construction reflect, but are much more severe than movements in the economy as a whole. There is no simple explanation of this phenomenon because in fact the pattern of change is the result of very different movements in the various sectors.

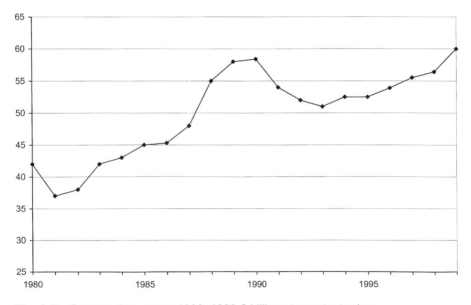

Fig. 2.5 Construction output 1980–1999 £ billion at constant prices.

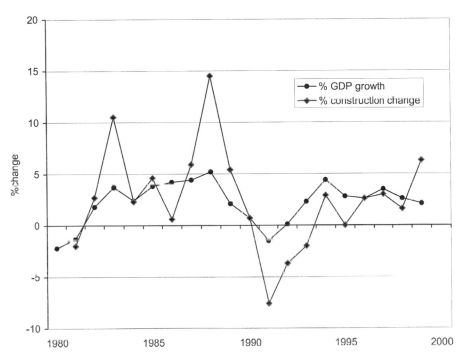

Fig. 2.6 % change in GDP compared with % change in construction output.

Figure 2.7 (a–d) shows new orders (not output) in four of these – public sector education, public sector health, private sector offices and, in the infrastructure category, new orders for railway construction. The first two are relatively stable, showing fluctuations within a general trend of overall growth. The first two are relatively stable though spending on health shows greater variation than spending on education. The difference between the two is accounted for entirely by shifts in government expenditure priorities.

Figures 2.5 and 2.6 showed a general slump in the period 1989–1991, but Fig. 2.7c. shows that this was particularly severe in office building – and if we were to look back further, we would see this sector has usually had the most severe switches in demand. These are clearly related to general economic conditions; as the economy booms, the demand for office space increases and vice versa. However the office market has its own dynamic something akin to what the American economic text books refer to as the 'hog cycle' because it was a characteristic of the Chicago pig market. In fact it is a characteristic of any market where there is a long time lag between the

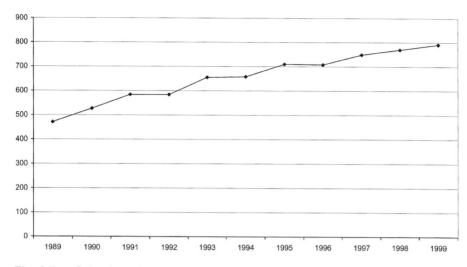

Fig. 2.7a   Schools and colleges – public sector (£ million).

Fig. 2.7b   Health – public sector (£ million).

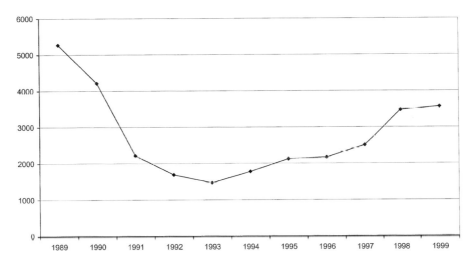

Fig. 2.7c　Offices – private sector (£ million).

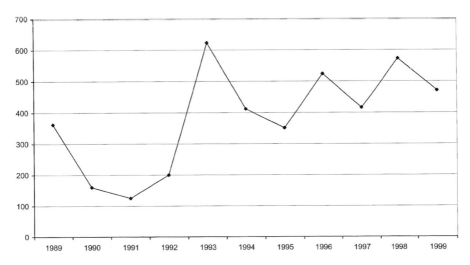

Fig. 2.7d　Railways (£ million).

start of the production process and completion. At times of high demand a shortage appears, the price of office space (or pigs) shoots up and output is increased to take advantage of the high profits available. But it takes time for the new output to come on to the market and when it does demand may have slackened. There is then a surplus of pigs – or offices – and prices fall quickly; production is cut back so that by the time of the next rise in demand there is again a shortage.

A full analysis of this and the whole complex interrelationship between government policy, economic cycles and the construction industry could fill a chapter or two. But perhaps enough has been said to give a fair indication of some of the factors involved.

Conclusion and summary

From all the evidence presented in the tables, charts and discussion above, some conclusion can be drawn about the significant characteristics of demand which might have impacts on the nature of the construction industry, the types and sizes of firms and their methods of working.

(1)   There is a huge range of different types and sizes of construction work, meeting a variety of needs, repairs to someone's front door at one end of the scale, to construction of massive office developments such as Canary Wharf.
(2)   The vast majority of jobs are quite small; though in value terms each is insignificant, in total they represent a large part of the industry's total output.
(3)   Repair and maintenance work represents nearly half the industry's total workload.
(4)   There are large fluctuations in the demand for construction over time and with particularly large variations in some sectors such as office building and housing.
(5)   Government policy, both its control of interest rates and its public expenditure programmes, has powerful effects on the industry's work.

To respond to a demand that is so varied and so variable over time an industry must have a very high degree of flexibility, and that the construction industry has proved to have – though often with some unfortunate consequences. How the structure of the industry can be interpreted as at least partly a logical response to the nature of demand is the subject of the following chapters.

For further study

The basic statistical information on construction orders and output is published now in *The Construction Statistics Annual*, replacing, from 2000, part of the *Housing Construction Statistics* which was published annually from the 1970s. It is accessible and downloadable from the Office for National Statistics website. It contains details of PFI schemes and government construction spending plans – which were not in the older publication.

A very useful source for information relevant to this and most of the chapters in this book is *Building's* 'Livenews and Archive' site www.building.co.uk. The archive section can be searched for relevant articles back to 1999, not only in *Building* but in its associated publications.

Modernising Construction (referred to earlier) gives a considerable insight into the attitudes and future practice of the Government as client. *Building's* regular economic survey, edited by Jackie Channon, tracks the overall demand for construction in a lively and readable format. *Public Private Partnerships* edited by Akintoye, Beck & Hardcastle discusses the actual practice of PPPs.

3 Profiles of the Industry

- Defining the construction industry – a problem of boundaries
- The structure of the industry – a statistical outline
- The construction business:
  - the majors: restructuring with continuity
  - the middle range: a brief review
  - the small operators: who are they and why so many?

Introduction: what is the construction industry?

We often speak of 'industries' as if they are single clearly defined entities – the car industry, the steel industry, the shipping industry and so on. In fact it is never easy to define an industry unambiguously; there is always a question as to what to include or exclude. Does the car industry, for example, include all the suppliers? Does it include the suppliers of the suppliers? Does it include the distribution network?

In this respect, construction is no different. There *is* an official definition available in the Standard Industrial Classification (SIC) but, official though it may be, it has boundaries and categories that do not seem to match today's reality. There are five major categories of work included: general construction and demolition; construction and repair of building; civil engineering; installation of fixtures and fittings (which includes plumbing, gas fitting, and electrical installation) and finally what is called 'building completion', that is painting, glazing, plastering, etc. The classification originally included open cast coal mining – which shows perhaps how arbitrary such classification can be.

However, this is not a particularly useful basis for analysing how the industry actually operates. For example the difference between house-building and other forms of construction, ignored in the SIC grouping, is a highly significant factor, as will become evident. Again, the exclusion of architectural and surveying practices from the official definition leaves out a very important group of actors in the construction process.

For the purposes of this book, then, we will take as a working definition of the industry *all those firms involved directly in the design and construction of buildings*. In much of the general discussion, we include civil engineering and infrastructure work, such as road building, bridges and railways, but the emphasis will be on building rather than civil engineering work. The materials industry, though obviously a fundamental part of the whole nexus of operations required to produce building, is excluded from the SIC construction division and it will not be discussed in any detail.

This chapter examines some of the information on the sectors of the industry engaged in actual construction, leaving the discussion of professional practices until Chapter 5. The first part analyses the industry's structure on the basis of the major available statistics while the second half looks in more detail at construction businesses and their areas of operation.

The structure of the industry

When economists talk about an industry's structure, they usually mean the distribution of firms by size (size is usually measured by output – or 'turnover'). This is because they have been interested mainly in the degree of 'concentration' – or the proportion of an industry's output in the hands of a few firms; the degree of concentration has in turn been of interest as an indicator of the strength or weakness of competition.

Some industries are highly concentrated; in the production of beer, cars, planes, ships, computers and dozens of others a high proportion of total output is in the hands of a few very large firms. This is true also of the production of many building materials – bricks, cement and glass for example; in the case of glass in the UK, output is virtually entirely in the hands of one organisation. Increasingly, in many industries concentration is occurring at an international level – as with cars, ships, aircraft and computers.

At the other extreme some industries consist almost entirely of very small organisations, especially local personal services such as hairdressing and dry cleaning. A third type of structure – which can be envisaged as a kind of pyramid – is one in which firm size is spread over a wide range from very large to very small, with a few at the top and a very large number at the bottom. This is the case with retailing for example; and as we will see in examining the statistics, it is quite clearly the case in construction.

The structure of most industries has not remained static; there has been a general trend to greater and greater concentration of output into the control of fewer and fewer firms. Even in areas where this was once thought to be unlikely, such as retailing, concentration has increased markedly as the big supermarket chains take over a larger share of more and more markets. It is not at all clear whether this is the case in construction but it is a question we will investigate in the course of this chapter – is the construction industry becoming more concentrated like so many others? If not, why not? These are not easy questions to answer.

Consider the current position first. The basic data are set out in Table 3.1 and derived from several different tables in *Construction Statistics*.[1]

Table 3.1   Number of firms, employment and work done by private contractors, 1999

Size of firm by number of employees	Number of firms	Total employment thousands	Value of work done £ million
1	88018	183.2	1471.1
2–3	49350	120.5	1364.8
4–7	16969	119.3	1173.7
8–13	4148	48.0	554.4
14–24	3271	63.2	773.2
25–34	1132	36.0	557.9
35–59	1188	58.1	1059.3
60–79	397	29.3	574.1
80–114	304	31.7	627.4
115–299	397	78.5	1405.2
300–599	105	47.3	868.6
600–1199	58	51.6	1099.3
1200 and over	42	92.3	1730.0

There are many problems of interpretation here. For example firms are categorised by their number of employees not by their turnover or output. This raises particular difficulties in construction because of the extent of subcontracting. The number of people actually employed by a construction firm will not indicate how many people actually work for it on the projects for which it is responsible; this would have to include all the subcontractors and *their* employees. The larger the project and the greater the scale of the firm's operations, the more likely this is to be true. A firm with relatively few employees may subcontract virtually the whole of its construction work. In terms of turnover (value of output) it could be very large, but in terms of employees relatively small.

There are other problems with the statistics, particularly what might be called the instability of the numbers of very small firms identified; some of these difficulties will be referred to later. Nevertheless these are the only available comprehensive figures for the industry as a whole and we have to make the best of them. Later on in this chapter different sources are used to classify the larger firms' *turnover* not *employment.*

The first column in Table 3.1 shows the size categories of firms by number of employees; there is certain arbitrariness about the categories, for if the figures are grouped differently they can give a different impression. But for now we stick to the published groupings. The second column of figures shows the number of firms on the register of the Department of Industry in 1999 in each of the size categories. The third shows the total number of people employed by firms in each category. The final column shows the value of work done by each group of firms in the last quarter of the year.

The most obvious characteristic of the industry stands out immediately from column 2, that is the very large number of very small firms, including 88 000 firms with only one employee; and this does *not* include an estimated 518 000 self-employed workers At the other end of the scale there were only 42 firms with over 1200 employees. There is another problem here in interpreting the figures. The top category of firms with over 1200 employees includes several firms which are very much bigger – Amec for example has over 20 000 employees and Skanska over 50 000; so a category of 'over 1200' does not tell us very much.

Column 3 shows the significance of these figures in a different way; nearly half the total number of employees in the industry (excluding, that is, the self-employed) were in firms of fewer than 13 employees; under 10% were employed in very large firms The experience of most people working in construction is of working for small firms.

When however we look at the final column a slightly different picture of 'significance' emerges. The large firms would be expected to be responsible for the higher value projects and a few of them would therefore be expected together to generate a higher value of output than a larger number of small firms; but the figures here still seem quite surprising. In the last quarter of 1999 the 42 largest firms did over £1.7 billion worth of work, more than the smallest 87 000 put together. The 100 largest firms produced work of the same value as the smallest 137 000. Nevertheless this does not show that the industry is in any conventional sense 'concentrated'. In fact these top 42 firms were responsible for only 13% of the industry's total output.

It has been argued that if we look at the statistics over the last twenty to thirty years, (see Fig. 3.1) we can see evidence of increasing concentration. The argument goes like this: if we compare in percentage terms the value of output produced by the firms with over 1200 employees, say in the two years 1972 and 1999, we can see that their share has increased. In 1972 the firms with over 1200 employees represented approximately 0.1% (*one thousandth*) of all firms and were responsible for 19% of output. In 1999 the

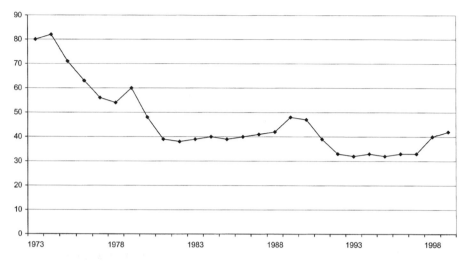

Fig. 3.1   Number of firms with more than 1200 employees 1973–1999.

number of large firms was much smaller; they represented only 0.02% of the total but were responsible for 13% of the output. This is proportionately a higher share. The proportion of large firms has dropped by 80% (from 0.1 to 0.02) while their share of output has dropped by only (6/19) = 31%. 'Concentration' it is argued, that is the proportion of the industry's output produced by the few largest businesses, has therefore increased.

However it is doubtful if the figures really justify this conclusion; and even if they did it does not seem very significant. There are many problems. For example, the highest group includes all those firms with over 1200 employees, and we do not know how that group is distributed; it could have been the case that the top 0.02 % in 1972 were also responsible for 13% of output, as in 1999.

Figure 3.2 certainly seems to show some sort of major structural change. The number of small firms increased and the number of large and medium-sized firms decreased. But there are many different possible interpretations of what actually happened. This was a period over which labour-only subcontracting (examined in detail in Chapter 4) grew substantially. It is quite possible that what the figures record is mainly a shift from direct employment by the large firms, rather than a reduction in their number. That is, many people formerly employed by the big firms were now operating in smaller groups as subcontractors. Again however, this interpretation is also speculative. There have no doubt been other factors at work; at some periods for example growth in small firms can be linked (statistically)

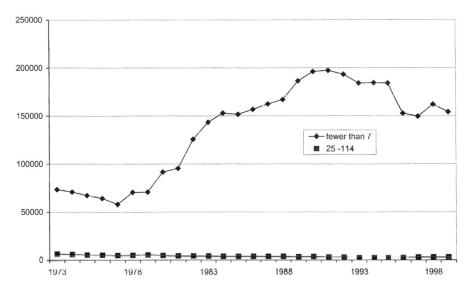

Fig. 3.2a   Number of firms with fewer than 7 and with 25–114 employees 1973–1999.

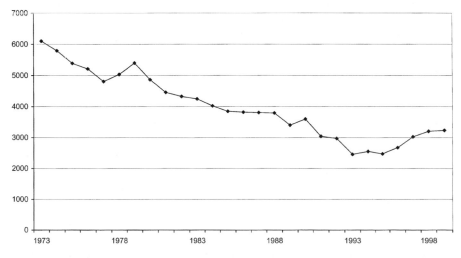

Fig. 3.2b   Number of firms with 25–114 employees 1973–1999.

with growth in repairs and maintenance expenditure – but in other periods there appears to be no such link.

The fact is that these statistics are not robust enough to allow us to draw firm conclusions about structural change. We perhaps have simply to accept them at face value. They show that construction has remained for forty years an industry with an exceptional number of small firms, but with firms of every size (in decreasing numbers) up to the very large. There was considerable growth in the number of firms with fewer than seven employees, peaking in 1990/1991, and a steady decline in firms of every other size group. Interpretation beyond that, *from these figures alone*, is problematic.

There is no doubt, from other evidence, that other structural changes have been taking place at many levels of the industry and we examine some of this evidence in the next section. However before moving on to that analysis, it is worth taking a brief look at another aspect of the industry's structure which is identifiable from the statistics themselves.

Although the conventional meaning of structure in economics is the distribution of firms by size, as discussed above, it is also possible to group firms by other characteristics – such as the type of work they specialise in. Table 3.2, derived from the detailed figures in Table 3.6 of the *Construction Statistics Annual*, shows the numbers of firms of each specialist type, this time in a very limited number of size categories. Even without the figures, incidentally, Table 3.2 gives a good representation of the nature of the industry as a whole for although it does not give a complete breakdown of all the trades (which would take several pages) it shows how the industry is structured in terms of the most important.

The figures call for little comment, except perhaps to point out that very few of the firms in the specialist trades have large numbers of employees. The only exceptions are electrical contractors and heating and ventilating engineering firms; the electrical contractors are the largest group of specialists in every size category. Most of these firms will usually be working as specialist subcontractors and of course modern complex buildings require the kind of high level expertise, combined with the capacity for handling big projects, which only larger firms will have.

The construction businesses

The statistical data looked at so far do give us a reasonable overall impression of the construction industry's structure and its change over time; but this kind of analysis is rather abstract and a better sense of the

Table 3.2 Number of firms by trade and number of employees

Trade	Small firms (1–13 employees)	Medium firms (14–19 employees)	Large firms (80–1200 employees)	Very large firms (>1200 employees)
General builders	38046	1735	330	8
Building & civil engineers	3040	409	139	12
Civil engineers	29240	471	104	18
Plumbers	12212	586	21	0
Carpenters/joiners	9828	177	11	0
Painters	8592	350	27	0
Glaziers	3000	113	12	0
Demolition contractors	718	67	8	0
Scaffolders	884	103	22	0
Reinforced concrete specialists	206	17	10	0
Heating & ventilating engineers	5075	375	42	2*
Electrical contractors	18474	814	93	6
Asphalt/tar sprayers	611	92	8	0
Plant hire	3906	155	21	0
Flooring specialists	2603	73	6	0
Construction engineers	781	52	4	0
Insulating specialists	821	100	13*	0
Suspended ceiling specialists	4496	22	10	0
Floor & wall tilers	747	14	8*	0
Roofers	5343	250	6	0
Plasterers	2454	80	7	0
Miscellaneous	17038	420	42	3*

*figures interpolated from different groupings in published statistics

reality of the industry can be gained from examining the construction businesses themselves and their activities. Obviously it is not possible to describe all the big firms in any detail, but a discussion of some of their characteristics and some of the changes that have taken place in the different sectors of the industry will help to put some flesh on the bare bones of the statistics. The sections below describe the industry in three main groups – first the giants, secondly the large to medium scale businesses, and thirdly the small firm sector.

The large firm sector

Building has for many years published a 'league table' of construction firms showing size by turnover and profitability. The positions at the top of the table in 1999 are shown in Table 3.3.[2] However this is a 'still' from a rapidly

Photos (3) & (4)   An industry of contrasts – from a man, a van and a ladder to tower cranes and steel structures.

Table 3.3 The top twenty housebuilders and contractors 1999

	Contracting Figures in £000	Housing Figures in £000	Property Figures in £000	Facilities management Figures in £000	Other Figures in £000	Total turnover Figures in £000
Amec	2 475 300		184 800	1 356 700	15 800	4 032 600
Bovis Lend Lease	3 931 034					3 931 034
Balfour Beatty	1 794 000			710 000	99 000	2 603 000
Taylor Woodrow	500 900	1 519 300	141 900		82 600	2 244 700
Carillion	1 129 400			614 200	249 300	1 992 900
George Wimpey		1 702 000				1 702 000
Persimmon		1 582 764				1 582 764
Laing	854 200	567 700	61 100		91 400	1 574 400
Mowlem	1 153 000		8 000	226 000	127 000	1 514 000
Skanska Cons/n	1 300 000					1 300 000
Barratt		1 259 500				1 259 500
Kier Group	937 400	97 400				1 034 800
Alfred McAlpine	347 075	463 405	10 057	131 451	2 050	981 038
Berkeley Group		868 948	73 541			947 489
HBG/Nuttall	929 300					929 300
Interserve	453 100	1 900		455 500	8 500	929 300
Wilson Bowden		568 800	158 000			726 800
Jarvis				725 779		725 779
Wilson Connolly		701 100				701 100
Amey				700 163		700 163

moving film; there has been much restructuring in the industry and the process of change may be accelerating. The league table positions of the major firms have shifted frequently. Some names that were consistently at the top have disappeared altogether; others have apparently appeared from nowhere (they were not simply lower down the league), while others have retained a position near the top over very many years. Constant change in actual positions at the top is not particularly surprising in an industry which is extremely competitive and in which the winning or losing of a single major contract (or disastrous losses on a contract once won) can change a firm's fortune very quickly indeed.

Nevertheless, in one sense there has been a remarkable degree of stability. Some of the apparent changes have partly been changes in name. For example, the top ten firms by turnover in 1994 were: Tarmac, Amec, P&O, BICC, Wimpey, Mowlem, Laing, Taylor Woodrow, Costain and Trafalgar House. Tarmac's building division is now Carillion, P&O were only in the league because of their ownership of Bovis (now Bovis Lend Lease), BICC begat Balfour Beatty, Trafalgar House's construction arm after several transformations finished up with Skanska. (These changes are discussed in more detail below.) So, substituting new names for old, it can be seen from Table 3.3 that in fact most of the same companies were still near the top of the league five years later, though Costain had slipped considerably and Laing may not be there much longer.

However, although the same core companies remain in the top group, most have been restructured several times in that period – and changes have seldom been changes in name only. Whether there has been more restructuring in recent years than previously, as has been claimed, is not easy to determine. It has been the case for a long time that as demand for construction as a whole or in individual sectors has fluctuated, large companies have moved in and out of different activities, seeking to increase profits or reduce losses. There have been periods of intense merger and takeover activity such as the late 1960s and early 1970s associated with high levels of demand for construction work in particular sectors (see Ball *Rebuilding Construction* Chapters 8 and 9 for detailed analysis of this period).[3] The mergers took place at every level of the industry among small, medium and large. At the top, major companies from outside construction moved in. So conglomerates such as Trafalgar House, P&O and Salveson acquired dominating positions, and construction firms diversified to virtually become conglomerates themselves; this was true for example of Wimpey and Tarmac.

In recent years again there seems to have been a frenzy of change, and changes in so many directions that the first impression is one of rather chaotic turmoil. Press descriptions of what was happening seemed often completely contradictory even within months. In May 1999 we were told that 'Conglomerates are out – the giant contractors were breaking up their empires'. Four months later there was talk of 'merger mania', a 'feeding frenzy' and a 'wave of consolidation'. But it is in fact possible to discern some logic to all this, which we attempt to do after examining some of the changes themselves, looking at a few specific examples.

Restructuring in the industry has taken the following major forms, which for the sake of identification later are numbered arbitrarily as types 1 to 6:

- Type 1 growth within one major construction area through acquisition, merger and internal development
- Type 2 diversification into related areas; the favoured fields in the 1970s were property and housebuilding (for contractors) or contracting (for housebuilders and civil engineers). More recently the moves have been into facilities management, services and consultant roles
- Type 3 diversification out of construction altogether – or diversification from outside into construction; this happened mainly in the boom years of the seventies and seems to have become less common. In fact the process has been largely reversed by:
- Type 4 consolidation on 'core' activities through de-merger, sell-off or by exchange of divisions
- Type 5 rationalisation by, for example, merging divisions within a company, reducing number of regional offices and nearly always by reducing employment
- Type 6 (really a subset of type 1 but worth identifying separately), takeover by or of foreign firms.

All these forms of restructuring have been taking place at the same time and sometimes within the same firms – which might seem a little odd as some of them are in opposite directions (consolidation and diversification for example). An explanation of these apparent contradictions requires an examination of some specific examples.

Amec

Amec has been until recently a clear example of type 1 – growth in one field (in this case engineering) through merger, acquisition and internal growth; but it is now also an example of type 2 (diversification into related fields).

The company was formed in 1982 from a merger of two of the UK's largest and most long established civil engineering groups Fairclough and William Press. Leonard Fairclough had set up a stone business in 1883 but by the beginning of the early 20th century had become mainly a bridge builder, constructing among others the transporter bridge across the Mersey and the Manchester Ship Canal at Runcorn – used until the new suspension bridge was built in the 1950s. William Press was founded in 1913 and became one of the country's major gas and oil pipeline engineers. AMEC acquired another large engineering group in 1988 – Mathew Hall, which had an even longer pedigree than Fairclough – going back to 1848, when it was set up as a lead working business. By the time they became part of AMEC all these companies had themselves made many acquisitions – mostly in closely related fields. For example Fairclough had acquired CV Buchan, a tunnelling specialist, in 1970 and Robert Watson and Co. – a leading structural steelwork fabricator; Mathew Hall had acquired the industrial buildings specialist IDC in 1985.

This sort of growth pattern is fairly typical of the major groups – particularly in civil engineering; and AMEC's story also illustrates another aspect of the modern industry – its growing international character. Overseas operations had been undertaken since at least the 1930s but it seems only to have been since the end of the 1980s that foreign groups have been acquired. In 1995 Morse International, an American company, became a wholly owned subsidiary; in 1997 AMEC took a 41.65% holding in the French company Spie Batignolles (whose history goes back to their early attempt to construct a machine to drill a tunnel under the English Channel in 1882) and in 2000 merged with AGRA, originally Canadian.

Restructured at the end of 1999 into three divisions, investments, capital projects and services (with a considerable loss of jobs), they moved away from competitively tendered, traditionally contracted work. Their chief executive, Peter Mason, explained the rationale in an interview with *Building*. Essentially the company has moved towards more certainty and lower risk – not into higher profit areas. Amec has 'balanced its exposure to the inherently volatile earnings from major projects with earnings from predictable activities such as long term work in utilities, rail, oil, gas and facilities management'. There will be less 'traditional hard bid contracting' and more negotiated partnered work and PFI schemes. To reduce the risk associated with the traditional contracting that they will continue to undertake, they have developed more sophisticated risk management techniques.[4]

P&O and Bovis Lend Lease

The story of P&O and Bovis has been one of frequent restructuring – with examples particularly of types 2, 3 and 4. P&O, originally the Pacific and Orient Steam Navigation Company (founded in 1827), had become a true conglomerate, having diversified into oil, property and many forms of international trading. It was its ownership of Bovis, taken over in 1974, which put it into the major contractors' league. Bovis's housing division was eventually floated off as Bovis Homes.

The contracting arm, as Bovis Construction, diversified to became a major international construction and project manager as well as contractor with over 60 offices in thirty countries and a staff of over 4000. In 1998, W.S. Atkins, Britain's largest engineering consultancy, was in talks to buy Bovis from P&O to create what was called at the time 'the world's leading one-stop design, construction and facilities management specialist'. The pros and cons were hotly debated in the financial and construction press, some commentators arguing that this was the way the industry had to go – an end to the old 'shipping to hotels' conglomerates and a move to more focused construction activities – but within that focus a wider capability – the 'one stop shop'. Then the deal collapsed. P&O announced their intention of keeping Bovis but expanding its operations. Bovis management said they were happy to stay with P&O. A few months after this (October 1999) Bovis was bought from P&O by the Australian company Lend Lease to become Bovis Lend Lease and P&O declared their intention to concentrate on their shipping (mainly ferry and cruise) interests though in fact they retained property and other interests as well.

Tarmac and Carillion

The story of Tarmac's origins in 1901 is well known. When Derbyshire's county surveyor Edgar Hooley noticed that tar leaked from a burst barrel and covered with waste slag from nearby iron furnaces had formed into a smooth hard surface he realised that this material could be exploited commercially and he set up a company, Tar Macadam (Purnell Hooley's Patent) Syndicate Ltd, to produce it. The business was taken over in 1905 by Sir Alfred Hickman, owner of a steelworks producing large amounts of hitherto useless slag. After operating as a materials producer and road builder for many years, the company moved into other areas through internal expansion and acquisition. It was probably at its most powerful and successful towards the end of the 1980s. The construction division was claimed to be the largest construction company in the country and had been responsible for major projects across the world; it had moved into house-building in the 1960s and with the purchase of John McLean in 1974 moved into the big league of housebuilders as well as contractors, second at one point only to Wimpey. Of course it was still the major producer of roadstone and other construction materials. So we see here major examples of diversification – but mainly within construction related fields, that is diversification classified as type 2 above.

From 1992 its fortunes wavered, despite its involvement in major contracts such as the Channel Tunnel. A loss of £350 million in 1992 was followed by the sell-off of many of its subsidiary businesses and con-centration on its three main activities – construction, quarry products and housebuilding. In 1996 it exchanged its housing division for Wimpey's construction and minerals, a move which resulted in major rationalisation on both sides, the closure of many offices and large-scale redundancies. Three years later the company was split into two, with the construction half being relaunched as Carillion, and shortly after, the rest of the company was sold to Anglo-American. This was consolidation on a grand scale (type 4 above) but also included elements of type 5 – rationalisation within the surviving divisions.

Although like Amec, Carillion is deliberately generating more of its income from PFI and services, it still has 80% of its turnover in traditional construction. In May 2001 it announced its intention to sell its social housing division – thus moving out of housing altogether and further consolidating in other forms of construction.

Trafalgar House and Skanska

Trafalgar House was, like P&O, a major conglomerate with interests which at one time included newspapers, a typing school, the Ritz Hotel and the QE2 as well as major housing and construction companies such as Ideal Homes and Trollope & Colls. In 1996 the company was taken over by the Swedish firm Kvaerner, which then divested itself of five operational areas (and some 25 000 staff), leaving only construction, process engineering, and an oil and gas division; again an example of consolidation and rationali-sation. Three years later Kvaerner sold the construction interest to Skanska and itself concentrated on its oil business. Skanska was (and is) the largest Swedish construction company with a workforce of over 50 000.

Balfour Beatty

The recent history of Balfour Beatty is unusual in that until the late 1990s it appeared to be the very junior subsidiary of BICC (British Insulated Calendar Cables) once the UK's main producer of electrical and communications cables – but it then in effect took its parent over. The construction division went through several forms of restructuring – selling its American contracting business and acquiring British Rail Maintenance for example. It worked on major projects with Amec including the new Hong Kong airport, Chek Lap Kok, the Heathrow Express and the Jubilee Line extension in London. In 1997 the company denied that there was any danger that it would be sold off by its parent BICC. Then in a remarkable turnaround it was the 'weak' cables business that was sold off, with BICC then actually becoming Balfour Beatty.

Laing

Some of the firms at the top of the rankings do not seem to have been subjected, until recently, to quite the same level of turbulence. Laing, for example, traces its history back to a small family firm started by James and Ann Laing around 1850 near Carlisle. Their grandson John William Laing, who joined the firm in 1894 at the age of 14, is considered to be the founder of the modern business. He developed it into a major international construction company covering everything from housing to large-scale engineering. Into its third century, the firm was still chaired by a Laing – Sir Martin grandson of John William. However the firm went through major rationalisation after making losses, mainly from its Cardiff Millennium Stadium contract. In 1999 it announced that it was to focus on private finance initiatives and prime contracts rather than traditional work; its turnover was to be cut by a third and its staff by 850 – a rationalisation which was forced on the company by major shareholders. In 2001 it was taken over by O'Rourke, a privately-owned company only a fraction of Laing's size. It seems to have been a rather sad and sudden fall of one of the country's great construction businesses.

Alfred McAlpine and Sir Robert McAlpine

The two McAlpine companies Sir Robert McAlpine and Alfred McAlpine are both descendants of the same family firm set up in 1878 by the first Robert McAlpine known apparently as 'Concrete Bob' (the first Bob the Builder?). His son Alfred decided to run his own business, based on the Manchester-based subsidiary of the parent company, in the late 1930s. Alfred McAlpine & Son was incorporated in 1940.

They are now very different companies. Sir Robert McAlpine, which does not appear in Table 3.3, has a turnover of about £390 million and is still privately-owned (though it was for a while a public company, a subsidiary of Newarthill) and largely controlled by descendants of the founder. Although in terms of turnover it is ranked about 35th in the league of contractors and housebuilders or about 17th among contractors only, it takes on major projects including in recent years the British Library, the Jubilee Line extension and the new Lloyds Register of Shipping building in London. It operates as civil engineer, builder and construction manager – but unlike most other companies until recently, does not treat each division as a separate entity. It still has outside interests – including a helicopter company – but these do not appear to be central to its operations.

Alfred McAlpine Group plc on the other hand is a public company – the last McAlpine family interest was sold in 1996. It was until recently a major housebuilder as well as a civil engineering contractor but sold the house-building division to Wimpey in August 2001. The reasons given for this sale in the company's press release show similar thinking to that of Amec quoted above. The company had acquired Kennedy, a leading supplier of utility services, in March 2001, an acquisition which with the sale of the housing division was part of a strategy to focus on infrastructure work – not just construction but every aspect; in the company's own words 'McAlpine will be focused on providing infrastructure services offering the ability to develop, design, build, own, operate and maintain the UK's infrastructure ... provide management services to utilities and highways ... offer design, local authority liaison, specialist asset replacement and maintenance, materials procurement and testing ... major customers include the big electric, gas and water companies.'[5] On top of all this, the company owns slate quarries in Wales and a quarry in New York.

The sell-off of Laing's construction arm and of Alfred McAlpine housing leaves only two firms in the top twenty with major interests in housing and contracting – Taylor Woodrow and Kier.

Taylor Woodrow

Taylor Woodrow, started as a housebuilding project (two houses and 100% profit!) in 1925, grew to become an important international engineering and building contractor. In the 1930s it diversified into merchant trading, building transport facilities in Africa and the Middle East. After the second world war it built some very big and innovative projects including Calder Hall, the world's first atomic power plant, and Hong Kong's Ocean Terminal. However more recently it has, like so many other firms, begun to consolidate and refocus, in this case into three areas – property, housing and contracting, though it seems to see its contracting now as mainly in support of its own property and housing development. It considerably increased its housing strength in 2001 with its purchase of Bryant Homes – which has made it the third biggest householder in terms of turnover, behind Wimpey and Persimmon. All the other housebuilders in the top group, with the exception of Kier whose housing branch is relatively small, are 'pure' housebuilders but in some cases have become so focused relatively recently.

Wimpey

Wimpey, a firm whose history, like those of Laing and McAlpine, goes back to the end of the nineteenth century, began as a quarrying and stone working firm. In the 1930s, it became the country's largest builder of council and private housing and has remained one of the largest housebuilders (by number of houses built) ever since, though it has occasionally been matched by Barratt and Tarmac and more recently was overtaken by Persimmon.

It diversified, however into virtually every form of construction. By the beginning of the 1990s it had a turnover of over £2 billion and was involved in major contracting, civil engineering, materials and property development all over the world. But in a remarkable swap deal it acquired Tarmac's housing division (McLean Homes) in exchange for its contracting business, consequently becoming almost solely a housebuilder. It operates through three divisions – Wimpey Homes, McLean Homes and Morrison Homes (working in the USA) – these in turn operating through 29 regional operating companies.

The latest major development, at the time of writing, has been the acquisition of the housebuilding division of Alfred McAlpine in August 2001, described above. Wimpey's recent history is therefore a clear case of consolidation (our type 4 restructuring). Each stage of consolidation has also involved rationalisation with job losses, including over four hundred after the McLean acquisition and the likelihood of more with its most recent purchase.

Beazer

In 1998 Beazer was the third biggest housebuilder (by turnover), and Persimmon the fifth; but in 2000 Persimmon took over Beazer to create a business which in terms of housing production outstripped even Wimpey and Barratt – a pure type 1 restructuring. It appears that this will be followed by a major rationalisation (type 5).

Barratt

Barratt has been almost purely a housebuilder for many years, though it was set up as a civil engineering firm in Newcastle and has from time to time diversified into other areas such as property development, but again seems to have focused back on its core activity. A much younger company than Wimpey, they grew extremely rapidly in the 1970s right through a period of recession in the market, mopping up smaller companies and their land banks, at first in the north and then throughout the whole country. Where the smaller firms had been unable to build profitably and were caught with land banks acquired on borrowed funds, Barratt succeeded by borrowing even more, by building very rapidly and very cheaply, but most of all by its energetic marketing of houses; it was, for example, the first housebuilder to mount large-scale national advertising campaigns on television.

Major firm restructuring – some conclusions

These examples could be extended to fill the whole book but perhaps enough has been said to enable some conclusions to be drawn about the general pattern of change among the major construction companies. In spite of the apparent contradictory directions of some of the movements there do appear to be a few clear trends:

● Consolidation into core businesses of housing or contracting – or specific types of contracting
● Shifts of focus towards services – facilities management and maintenance
● Consolidation of the housing sector through a series of takeovers. (There have been many more recently than the examples given above.) In fact housebuilding is showing clear signs of moving to real concentration in the way that manufacturing industry has done. It will be interesting to see how far this process can go.

There seem to have been a number of different market factors driving these changes. One of these has been the Private Finance Initiative; a number of recent takeovers have had as one major objective access to the expertise required to bid for and carry through PFI projects; this seems to have been true of Skanska's purchase of Kvaerner's construction business. As we saw in some of the case studies above, diversification by firms into facilities and services has had the same motive.

A second market characteristic has been the very large scale of projects coming forward which have often been too big for even the largest firms, who have therefore tendered as consortia. The Channel Tunnel was the obvious example but there are also airport developments, shopping centres and the huge office developments in London's Docklands. The vast resources required for such projects mean that size itself is an advantage but also that the development of specialist skills gives a company the ability to participate in major joint ventures. An interesting example here was the joint bid by the two McAlpine companies, who had not worked together for 60 years, to develop the Eden Project in Cornwall; the idea was to combine Alfred McAlpine civil engineering expertise with Sir Robert McAlpine building experience as well as to spread risk.

Undoubtedly another influence on restructuring has been the increasing influence of major shareholders and city analysts. That pressure came through very loud and clear in discussions of many of the internal rationalisations and mergers that took place at the end of the 1990s. The City's views, insofar as they were consistent, were that contracting alone did not yield sufficient profit margins and was too risky; that a link with facilities management and service provision gave more security but that trying to keep too many diverse activities under one management could lead to disaster. (In earlier years the City had supported conglomerate structure as a guarantee of security on the grounds that if part of the business failed, the rest could maintain viability.) There was pressure to focus on 'core business'. Before the division of his company described above, Sir Neville Simms, chairman of Tarmac, was quoted as being quite explicit about giving in to City demands: 'If the City won't rate keeping my business together it's not my job to say I'm right . . . I have to turn over every stone to lift shareholder value'.[6]

One interesting aspect of these shareholder pressures is that profit levels alone are not the most significant factor. There is in fact conflicting evidence about the profitability gains from merger, certainly in contracting. In housebuilding however, merger followed by rationalisations of the sort undertaken by Wimpey and Persimmon does have increased profitability as a major objective. In general, profitability in housebuilding is currently at about 8–10% compared with around 2–3% in contracting; this might have been expected, other things being equal, to stimulate a rush into housebuilding. However, without the necessary experience and expertise, that would be a high risk strategy; what is looked for in contracting is long-term growth prospects and security rather than high profit with high risk – hence the supporting of traditional work with activities such as facilities management.

Not all firms have restructured in the particular ways described above and there is no agreement on whether the trends will intensify or whether the current phase of restructuring is coming to a close. What does seem certain is that there will continue to be change of one sort or another. More takeovers by large foreign businesses such as Skanska seem particularly likely, with giant firms such as the French Bouyges already having shown some interest in acquiring a stronger foothold in Britain. Similarly the German firm Hochtief has been reported to have interests in buying Balfour Beatty or Carillion (or both – that would really re-structure the industry at the top).

The middle ranks

There is no real break in size (by turnover) between what might be called medium-sized companies and the major firms discussed above; in fact, after a large gap at the top between the £2 billion of Amec and the £1 billion of Carillion, the turnover figures of *Building*'s top 100 firms in 1999 were fairly evenly distributed down to £70 million. The real distinction to be made is in area of operation with the medium-sized firms working mainly in one or a limited number of regions while the big firms work across the country (as well, of course, as abroad). The situation is never static however – as many 'regionals' have become national by a steady programme of acquisition. Below the £200 million turnover level most firms are regionally based. A definition of medium as firms with turnover between about £30 million and 200 million seems reasonable though it is bound to be arbitrary.

This range includes a very large number of firms whose activities, in aggregate, include every aspect of construction as well as much else besides so any generalisations or examples are unlikely to be wholly representative. The middle range businesses fall roughly into the same categories as the major companies discussed above and are subject to many of the same pressures: fierce competition, threats of takeover and for the public companies, continual pressure from shareholders to increase profits and improve shareholder value. More of them are, however, privately owned and the range also includes the larger of the specialist subcontractors and firms providing building services.

There are many speculative housebuilders in this medium size range. It is difficult to be certain without more detailed research but one of the major changes in this group over the last thirty years has been a reduction of numbers in some regions but remarkable stability in others over quite a long period of time. Of 15 regional housebuilders working mainly in the southern region in the 1970s eight are still going strong. Of the fewer firms in the northern region, none appear to exist in the same form; some grew and were taken over by larger concerns (such as Whelmar, Broseley and Northwest Holst, now French owned and a contractor rather than a housebuilder). Some expanded very fast and collapsed even more rapidly (Northern Developments). Others have disappeared without trace. However successful new firms have been established in the region and have grown rapidly to operate over a wide area, Redrow being the best-known example.

Often the medium to small housebuilding firms survived by building at the top end of the market where scale economies are less important and margins on individual properties are higher. When demand was less buoyant the fierce competition made life very difficult for the smaller firms. In the north west fierce competition from Barratt in particular drove many out of business. What also seems to have happened is that conglomerates took over housebuilders in times of boom, perhaps mainly for the increasing asset value of their land banks, but very quickly lost interest in the housebuilding operation itself when the markets slumped. This certainly seems to have been the case with the takeover of a successful regional builder, Whelmar, by the conglomerate Salveson.

Most middle-range firms are, like most of the majors, specialised in one field. There are some exceptions and it will be interesting to see if they can maintain their spread in view of the current market pressures. One example out of very many is David McLean Holdings, founded only in 1972 as a contractor but moving through work for housing associations into private housebuilding. It is now two separate businesses, David McLean Homes building about 400 houses a year and David McLean Contractors working through five divisions to produce a turnover of about £90 million a year. It appears to be a dynamic company, embracing all the new post-Egan approaches; it has framework agreements with some prestigious clients including the HBOS Bank and has developed partnering agreements with a group of subcontractors. It also has a multi-disciplinary design division recognised as an architectural practice by the RIBA.

This may not be a typical medium-sized firm but it shows that firms of that size can achieve the spread of activity and perhaps styles of management of their much larger contemporaries. Whether the production of such a small number of houses can keep the housing business viable remains to be seen.

The small firm sector

Why so many?

As we saw earlier (Table 3.1) the vast majority of firms are small and most are very small indeed. The reasons for the existence of so many small operators are fairly obvious and are unlikely to change. They are the same today as those identified in a government-supported study of small building firms carried out by Patricia Hillebrandt as long ago as 1971.[7]

First there is the *type of demand*, some aspects of which were described in the last chapter: the fact that so much work required is small-scale repair and maintenance, which often can be carried out even by a man working on his own. This link between the number of small firms and the demand for maintenance and repairs, particularly in housing, comes out very clearly in the statistics for output analysed by size of firm. Firms with fewer than seven people do 63% of housing repairs. If we include as small all firms with fewer than 114 employees the proportion rises to over 90% (Fig. 3.3). Only one per cent is done by big firms – and these will certainly be large local authority or housing association contracts. The position with other forms of maintenance and repairs is slightly different, but still a quarter of the work is done by these very small firms.

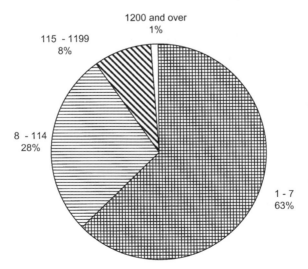

Fig. 3.3a   Housing repair and maintenance.

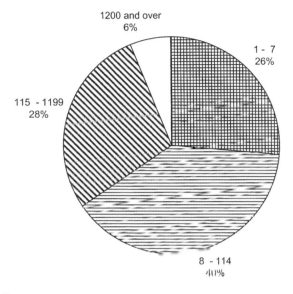

Fig. 3.3b Other repair and maintenance.

Many small firms are also involved in building new work at the smallest
scale such as single or small groups of houses, development of small
commercial sites, farm buildings, house extensions and so on. Both the
small-scale maintenance and the small-scale development are usually
thought not to be worth while for large firms as they cannot achieve the
scale economies available on bigger projects. But this is not universally true;
in fact the major contractor Mowlem has set up a small-scale maintenance
arm Skillbase to undertake very small-scale maintenance work; it is not
clear yet how successful this is but it is unlikely to become a serious threat to
good local small-scale operators.

A second major reason for the significance of small firms in the industry is
the fact that the *demand has to be met where it arises*. Like consumer demand
for most things, demand for buildings is spread geographically across the
whole country; but whereas new cars, football boots or garden gnomes can
be delivered quickly from anywhere in the world, buildings – at least,
conventionally-built buildings – have to be manufactured on the spot. The
local builder is in a position to know about local demand and local clients
requiring small-scale work will usually contact local builders. This last
point (local clients) is treated as a separate reason in the 1971 Hillebrandt
report – the nature of the client; but it is really part of the same feature of
construction – its dependence on specific locality.

A third reason is the *importance of subcontracting*. Building has since the early nineteenth century been carried out by a main contractor working often with large numbers of subcontractors; and although on big projects the subcontractors themselves may be moderately-sized firms, the widespread practice of labour-only subcontracting means that single tradesmen or small groups of tradesmen, each of whom may be classified as self-employed, can hire themselves out to work for others. This is one of the most common ways in which many individual tradesmen and small groups, such as bricklayers, work. There are many significant and highly contentious issues involved in subcontracting of this sort which will be discussed in much more detail in the next chapter.

Finally there is what economists call the *'low costs of entry'* into the industry. It is possible to set up in business with virtually no capital. The tools needed for most repair work are relatively simple and cheap; a secondhand van will do for transport and it is easy to hire any more expensive equipment needed on a short-term basis. There is no need for much working capital in the form of materials stocks. Most householders who have had small building jobs done will have come across the builder who asks for an advance to buy the necessary materials (or even a loan to buy his lunch!) and it is common practice for builders merchants to extend limited credit to local contractors. Indeed the large network of builders merchants, again ranging from very large to very small firms, represents a resource for small builders without which they could hardly survive.

Types of 'small firms'

The category of small firms – as with the medium – covers of course a range of types of business and types of work, from the man (and in construction it is almost always a man) working on his own doing general maintenance to the skilled specialists usually working as subcontractors. Again, the categories as well as the reasons identified in the Hillebrandt report are still applicable (which is perhaps an indicator of how little changes and how slowly in some sectors of construction). That report identified five types of small builder:

- general builders doing mostly repair and maintenance with some new build
- small general builders specialising in housebuilding
- small civil engineering firms
- specialist subcontractors
- builders with very few employees but who carry out large contracts

through subcontracting (if size were defined by turnover rather than employees then this last category would probably not be considered small – the same point made above in the discussion of the statistics).

But then there are further distinctions to be made. Even among the single ownership general contractors there is great variety; many are trained in specific skills but take on all forms of work, perhaps calling in assistance from other tradesmen when needed as well as employing general labourers on a casual basis. This is actually the oldest form of working in the industry and is mentioned again in the next chapter. Others are essentially odd job men who have acquired their skills by initially working with others – or at the worst – through trial and error.

The group identified by Hillebrandt as general contractors specialising in housebuilding also includes different sorts of operators. A detailed study[8] carried out of the small housebuilding sector on Merseyside in the 1970s found there were three basic types; they were classified as the financial speculator, the building contractor who built the odd house or small group of houses when a profitable opportunity occurred and the small housebuilder who would do some contracting when the housing market weakened. The first type were not really builders at all; in one case studied the directors had interests in ballrooms, trouser pressing, 'wrecking' and travel agencies. At peaks of the housing boom this type of speculator believed easy profits were to be made, would buy up a plot or two of land and subcontract all the building work to local contractors. Occasionally they were successful; as often as not they were soon in financial difficulty.

The second group – contractors who occasionally built houses – were extremely common; potential profits from a single project always seemed attractive compared with the low returns on the competitively won contract. But again without a real understanding of the land and housing market such ventures could and often did end in disaster. Many of the firms that prospered, though, found a new source of less risky work in contracting to build houses for the small housing associations that sprang up in the 1970s.

The third group, the real housebuilders, included many small strong family firms with long histories of surviving on a very modest output; they usually had few or no permanent staff but regularly used the same subcontractors. Others in this group came and went within a very short time. In one period, fewer than half the housebuilders registered with the NHBC in the region lasted for ten years while some had been building houses in the locality for over 50 years. How typical this was of the country as a whole and how far the situation changed by the end of the century is difficult to say without similar detailed research. But it does seem that in some areas the small housebuilder is now rare, one reason almost certainly being the increased ability of the larger firms to operate successfully on small sites.

Conclusion

This chapter has outlined the overall structure of the construction industry through an examination of the official statistics and through a brief discussion of some individual firms and sectors. Despite the industry's complexity it is possible to make some reasonable generalisations about the present position and perhaps it is also possible to identify some developing trends.

There is obviously a considerable amount of restructuring going on but whether this amounts to increasing dominance by the largest firms is difficult to say in the case of contracting. There always have been giant firms at the top (the former Tarmac for example). It seems more that they have changed their character. The old conglomerates have gone and been replaced by more focused organisations which still, however, combine two or three basic forms of operation, for example contract management and services as well as contracting. However, the possibility of more mergers and in particular takeover by giant European firms is very real and could alter the picture radically.

In the case of housing, the situation is very interesting because it does seem that concentration in the classical economic sense may be taking place. More possible mergers and takeovers are being predicted as this is being written. There has certainly been more concentration on housebuilding as a separate and different activity from contracting.

However, in the past, as economic conditions have changed, the fashion for different forms of organisation has also changed. It is not totally inconceivable that if we see recession again, we might also see a new move back to diversification as a form of spreading risk.

For further study

The *Building* website www.building.org.uk recommended at the end of Chapter 2 is also an extremely useful source for updated information on construction firms. The other major source is of course the websites of the individual companies. These vary enormously in quality and usefulness – some giving company histories, access to reports and accounts, descriptions of recent projects, etc., others saying very little indeed. Particularly good ones at the time of writing are: Galliford www.galliford.co.uk, Alfred McAlpine www.alfred-mcalpine.com, Carillion www.carillionplc.com, Kvaerner www.kvaerner.com.

The economist Patricia Hillebrandt has produced many studies of the industry including *Construction Companies in and out of Recession* (1995) and with Jackie Cannon, *The Modern Construction Firm* (1990), both published by Macmillan. The 1995 book gives a detailed analysis of company finances and strategies during the recession of the early 1990s, revealing the effect of recession and how several of the major contractors responded by restructuring in various ways.

4 The Workforce on Site

- The industry is its people
- The evolution of industrial relations: from guilds to unions; through conflict to negotiation
- The structure of employment – employed, self-employed, sub-contracted; taxed and untaxed
- Training and skills – the enduring problems
- A dangerous industry – but why?

Introduction: a commitment to people?

However advanced the technology an industry employs, its effectiveness depends ultimately on the people who run it. In an industry like construction, which despite the modern techniques is still very labour intensive, the skills, commitment and effective organisation of the workforce are overwhelmingly important in determining whether the industry can produce a high quality of output at a fair price. In fact it could be argued improvements in productivity in construction are as likely to come from improvement in employment practices as from advances in technology.

It is a point the Egan report underlines by making 'Commitment to People' one of its key drivers of change:

> 'Commitment to people means not only decent site conditions, fair wages and care for the health and safety of the workforce. It means a commitment to training and development of committed and highly capable managers and supervisors. It also means respect for all participants in the process... It is the Taskforce's view that construction does not yet recognise that its people are its greatest asset and treat them as such'.[1]

The last sentence in that quotation is a remarkable indictment of an industry which has existed in its modern form for two centuries and in which there has been a long series of attempts to improve labour relations and working conditions. If it is a fair criticism it needs some explanation.

This and the following chapter look at some of the basic facts of employment in the industry, some of the historical developments and some of the persistent issues and controversies in an attempt to give a clear picture and explanation of the current situation. There are at present roughly one and a half million people working in construction, that is approximately 8% of the working population. That figure would be very much higher if we were to include all those who though not directly involved in the production of buildings and other construction work, are closely connected with the industry in other ways – the staff of the many legal firms specialising in construction, the civil servants in the relevant departments of central and local government, housing association staff, the researchers engaged in research establishments, the employees in the building materials industry and so on.

These two chapters examine some characteristics of the construction workforce itself – the people physically involved with the actual production of buildings and other construction works. Chapter 5 will discuss the roles and organisation of consultants such as architects, surveyors and engineers and the construction managers themselves. This chapter looks at different aspects of employment of the people who have been traditionally described as 'operatives' – the skilled tradesmen and others actually working on sites.

The chapter begins with an outline of the development of modern forms of employment and of labour relations; as in other areas, the history helps to explain some of the rather curious characteristics of employment in the industry today. The next section examines some recent changes in the structure of employment and looks in detail at the important phenomena of self-employment and labour-only subcontracting. The high level of these ways of working differentiates construction from virtually every other industry and is a major contributor to many of the weaknesses for which the industry has been constantly criticised such as the poor safety record, the inadequacy of training and consequent skills shortages.

Later sections look at these two issues, safety and training, which have for so long and so frequently appeared in the headlines of the building press as sources of concern.

The evolution of industrial relations

Early development – the guild system

Building workers always seemed to have felt themselves vulnerable to exploitation – by clients and by employers and indeed by other building workers undercutting their wages or debasing their skills; clients and employers on the other hand have always seemed equally to feel themselves vulnerable to exploitation by the workers – through overcharging, shoddy work and failure to fulfil contracts or agreements. In the 1400s workmen were fined for 'being late, for idling, quarrelling, losing their tools and obstructing other workmen'. But they in turn had their methods: 'they conspire together that no men of their craft shall take less for a day than they fix and that none of them shall do good steady work which might interfere with other men of the craft . . .'.[2]

This apparently permanent mutual suspicion and antagonism – of worker against client and client against workmen and also one group of tradesmen against others – has throughout the centuries been kept under some kind of control by rules, procedures, traditions and regulations of many sorts.

In the middle ages the system of craft guilds developed which controlled the training of craftsmen, the acceptable rates of pay and the standards of workmanship. The guilds, which were based on the type of materials which their members worked, existed in every town but were often organised across wider areas of the country. There was a hierarchy of workers: the apprentices at the bottom, learning their trade for at least seven years, the journeymen – whom we would now call the qualified craftsmen – and the masters, who could themselves take on apprentices. The masters were usually in charge of the parts of jobs needing their particular skills – say the timber framing for a merchant's new house.

In later periods (after the Statute of Artificers of 1563) and in many areas of the country, wage rates were determined by the local magistrates, often in agreement with the guilds – but the guilds themselves continued to control the apprenticeship system and held themselves responsible for high levels of craftsmanship. The system was in many ways a very satisfactory arrangement for both sides – the guilds guaranteed quality in return for a fair wage. Of course, a fair wage was often barely above subsistence level and a day's work could be from five in the morning to ten at night.

The guild system survived for centuries, though obviously going through many changes. It decayed and ultimately disappeared as direct employment by the new breed of builders and contractors began to overtake in importance the old system of working directly for clients. This development will be described in more detail in Chapter 6; it was all part of the general shift in working patterns brought about by the industrial revolution.

Photo (5) Membership certificate from the Amalgamated Society of Carpenters and Joiners – late nineteenth century.

Trade unions arrive: strikes and lockouts

In the late eighteenth and early nineteenth centuries most trades had local clubs and friendly societies for mutual support, which, like the old guilds from which they were descended, took on the responsibilities for helping unemployed members, giving financial support for widows and children, as well as seeking to maintain craft standards.

Although the Combination Acts of 1799 and 1800 made it a criminal offence for workers to combine against employers to gain increases in wages or reductions in hours, building workers managed to find ways often through links between these small local organisations (and often at the cost of fines and imprisonment) to put pressure on employers; so much so, that only two years after the passing of the Acts, employers were complaining about workers holding the whip hand. The Mayor of Leeds complained that

'Perquisites, privileges, modes of labour, rates, who shall be employed, etc., are now all dependent upon the fiats of our workmen, beyond all appeal.'[3]

This complaint was to be heard repeatedly throughout the next two hundred years and still is (see for example press accounts of the electrician's strike during the building of the Jubilee Line extension in London). The complaint was sometimes justified, but in the end power remained with the employers who won most of the battles during the nineteenth century.

It was only after the repeal of the Combination Acts in 1824 that real trade unions in the modern sense could come into being. At first these were mainly local groups fighting local battles, usually over wages, hours and aggressive – or often brutal – foremen, but the workers were well aware that they would need national organisation to achieve real improvements in conditions. In the building industry it was the carpenters and joiners, who had for long been among the better organised tradesmen, who formed the first national building union, the Friendly Society of Operative House Carpenters and Joiners of Great Britain. Among the aims were

'to unite in bonds of friendship for the amelioration of the evils besetting our trade; the advancement of the rights and privileges of labour, the cultivation of brotherly affection and mutual regard for each other's welfare.'

The high moral tone and the real idealism were common – and still impressive – characteristics of all the early building unions.

Within a few years, the stonemasons, plumbers and bricklayers had all formed unions and in 1832 the first attempt was made to form a national organisation incorporating all the trades – the operative Builders Union; but it lasted only a year before over ambition, external opposition and internal squabbles led to its collapse.

The long history of struggle that followed, conflict between 'masters and men' as the Victorians described them, conflict within and between the unions themselves, their successes and failures has been told elsewhere.[4] It is possible here to summarise only briefly the main lines of development and to identify some of the persistent themes and sources of conflict, most of which have continued through to the present day.

The story is essentially one of long and faltering progression towards greater amalgamation of unions in a search for greater strength and unity; it is a story of strikes and of lockouts but also of repeated attempts to find common cause with employers and establish national forums for negotiation on working conditions.

One of the early big disputes that showed growing national solidarity was the stonemasons' strike at the Houses of Parliament in 1841 referred to in Chapter 1; the strike initially started over the behaviour of the foreman, George Allen, who, said the workers representatives, 'damns, blasts and curses at every turn'. The employer admitted Allen was 'a rough diamond, not particular in his expressions, but then a man who has received a university education was not adapted for a foreman of masons'. The significance of the battle that followed was that it took on a national dimension when the contractors sacked the masons and tried to bring in substitutes from other parts of the country. Eventually the unions lost and were considerably weakened.

But the battle lines between unions and employers were now clearly drawn. The areas of dispute were of course wages and hours of work – but there were also frequent disputes over payment methods; the unions maintained a consistent opposition to piecework and hourly rather that weekly payment. Employment of subcontract, particularly non-union, labour, was a constant source of friction as was the control of apprenticeships. Most unions were keen to keep the apprenticeship system but wanted the numbers of apprentices limited so that in times of slump there would not be too great a surplus – leading to falls in wages – and in times of boom, wages would rise rapidly. Employers on the other hand were concerned even then at the shortages of skilled labour and wanted all restrictions on apprentice numbers removed.

The first major conflict that involved unions from all trades together with the labourers started with a demand for a nine-hour day by carpenters and joiners in London – put in the most respectful terms:

'Gentlemen we the men in your employ consider that the time has arrived when some alteration on the hours of labour is necessary; and having determined that the reduction of the present working day to nine hours at the present rate of wages, asked for by the building trades during a public agitation of eighteen months, would meet our present requirement, we respectfully solicit you to consider nine hours as a day's work. A definite answer to our request is solicited by the 22nd July 1859.'[5]

The first response of one of the major employers was to sack the men who presented the petition. There followed a lockout by all builders with more than fifty employees; it lasted over six months putting 24 000 operatives out of work. Employers tried to force workers to sign what was known as the Document, agreeing that they would not in future join any trade union. Ultimately the union lost over the nine-hour day; to make matters worse the employers now brought in hourly instead of weekly wages and extended the use of piecework – both of which were bitterly opposed by the unions. But they won on their refusal to accept the Document.

After a period of relative failure, the unions began to regroup, growing in strength and organisation. The Amalgamated Society of Carpenters and Joiners (ASCJ), formed in 1860, had by 1870 under the leadership of a very remarkable man Robert Applegarth grown to a membership of more than 10 000 with over 230 branches round the country.

By 1875, it had over 25 000 members. Other unions had also become stronger – the Stonemasons with over 11 000, the Bricklayers Society 6000. Many were still very small and total union membership was probably just over 100 000, still a tiny proportion of the total workforce of 875 000.

The last two decades of the nineteenth century were a period of boom in building; unemployment was low and the unions grew stronger, with the largest – the ASCJ – reaching a membership of 65 000 or about a quarter of all carpenters and joiners. After many failures, the labourers too managed to establish a successful union – the United Building Labourers Union.

As unions had developed in the nineteenth century the employers saw it in their interest to form associations to fight jointly against what they saw usually as the unreasonable demands of the men. The National Association of Master Builders (later the National Federation of Building Trades Employers) had over 1000 affiliated members by the beginning of the twentieth century. The struggle between the two sides continued at times with considerable ferocity and bitterness. Employers still protested at what they saw as the unions' bullying. But for the thousands of workers inside as well as outside the unions life was still hard; there was no job security, pay was still low – though it improved in boom periods; there were no insurance or pensions or compensation for injury.

Towards national negotiation

Despite the continuing disputes, something of real importance had been achieved by the turn of the century, namely the growing recognition that the way forward was through joint negotiation. There were still plenty of employers who thought unions should be banned and still plenty of union members who believed the whole building industry should be nationalised and run as a grand workers' cooperative. But the realists were working hard on practical ways forward. There were many false starts as for example, the setting up of the so called Builders' Parliament which was an attempt to establish a joint employer–employee council. It was officially renamed the Industrial Council for the Building Industry in 1917 to fit in with a government scheme (the Whitley Councils); in fact the building industry was the only one to respond so positively to the government's attempt to establish such bodies in many industries. It started with high hopes:

> 'the interests of employers and employed are in some aspects opposed but they have a common interest in promoting the efficiency and status of the service in which they are engaged and in advancing the well being of its personnel.'[6]

The council collapsed when the employers withdrew in 1923 – but the unions' own organisation has been strengthened by the creation of the National Federation of Building Trades Operatives (NFBTO) in 1918. The employers had formed the National Federation of Building Trades Employers in 1899 so the basis was there for national negotiation. Progress was not easy though. There were continuing internal disputes; the Liverpool employers, for example, remained outside the NFBTE until 1940. However, again after a few false starts, the machinery for negotiation was established in the National Joint Council for the Building Industry in 1926.

The agreement and rules of the NJC occupied many pages but two clauses (in the 1963 version) indicate the crucial function:

> 'Clause I: It is agreed the rates of Wages of workmen employed in the Building Industry, the hours of labour and other such matters ... shall be determined on a national basis.

> Clause 5: The duties of the Council shall be:
> To deal ... with the following matters: i. Rates of wages; ii Working hours; iii. Extra Payments; iv. Overtime; v. Night gangs; vi. Travelling and Lodging Allowances; vii. Guaranteed Payments in relation to Time Lost; viii. Termination of Employment; ix. Apprenticeship; x. Holiday payments; xi. Safety, Health and Welfare conditions.'[7]

The results of negotiations were codified in what was known as the Working Rule Agreement – which though revised virtually every year, still survives. The National Joint Council has now been replaced by the Construction Industry Council; the union members are UCATT, TGWU and the GMB and the employers are represented by the Construction Federation, the National Federation of Roofing Contractors and the National Association of Shopfitters. The AEEU, Association of Electrical and Engineering Unions, is not part of this agreement and negotiates separately.

As late as the mid 1990s the Working Rule Agreement was very much the same document as it had been in the early days – it ran to over 160 pages. It was felt by many in the industry to be rather antiquated in format and wording and indeed there was pressure for it to be abandoned. The current edition is a very much reduced version issued in two parts and setting out a three-year agreement. The main document sets out the conditions of employment, for example, working hours, holiday entitlement, sick pay and union rights. The supplementary document of only eight pages lays down basic rates of pay and allowances for the various grades of worker, for example, the general operative rate starting from 24th June 2002 is £5.49 per hour and the top craft rate is £7.30 per hour. These rates, which are minima, will be considerably exceeded on many sites but the statistics for earnings show that on average the WRA rates represent only slightly less than actual payments. Hourly earnings excluding overtime were £7.35 in 1999, very close to the average of all industries – £7.36. Gross weekly earnings for male manual workers were £351 per week compared with an average over all industries of £335.

So in theory there is a firm basis of agreement between employer and employees, which should give stability and eliminate conflict. It does not, of course; the three-year pay agreement of 1997 was reached only after much argument and some serious strike threats.

Again, the Working Rule Agreement can be strictly enforced only where the companies and employees are members of organisations that accept it. In the case of the employees this is in fact a small minority of the total. Though the unions have successfully argued in the courts that where men work on a site run by a contractor who is signatory to the agreement, they are entitled to the rights agreed under the WRA, there is often in fact a serious obstacle and that is the very high proportion of people who are, nominally, self-employed. This is an issue of great importance in the industry and is examined in the following section.

Union membership fluctuated throughout the twentieth century, with a rise immediately after World War I, a sharp decline in the depression years, growth in the late 1930s with further growth and considerable political strength under the post World War II Labour government. After 1970 membership declined – leading to more union amalgamations. Today there are essentially four unions representing building and civil engineering operatives' interests – UCATT, GMB, TGWU and the AEEU. Union density, as it is called, that is the proportion of union members on building sites, is reckoned to be about 15%.

The structure of employment

Total employment in the industry has fluctuated more or less directly with construction output at least up to the mid 1990s. If Fig. 4.1 is compared with

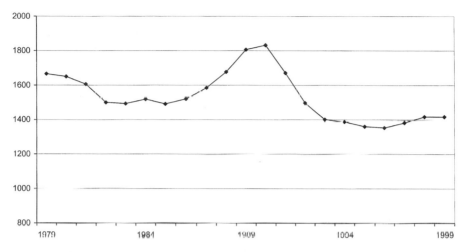

Fig. 4.1   Construction manpower 1979–1999 (thousands).

Fig. 2.5 in Chapter 2 the similarity of the overall patterns is clear, both output and employment peak in 1991. After 1995 however the steep rise in output is not accompanied by a steep rise in employment. Labour productivity (output per person) seems therefore to have increased quite markedly. It is unwise to draw quick conclusions from figures like these without a careful study of underlying factors; the improvement may be due, for example, to an increase in efficiency on site or may be the result of differential changes in construction prices, values and wages. Interesting though they are we leave those issues aside here.

The overall pattern of change in total employment shown in Fig 4.1, which includes administrative and professional employees, is not surprising. However, Fig. 4.2 tells a very different and remarkable story – or rather several stories. We can see there a transformation of the structure of the construction workforce over a relatively short period.

First, looking at the lowest graph line, we can see the decimation of the local authority direct labour force. There are a number of reasons for this, first there was the decline in local authority housebuilding, secondly it has been the deliberate policy of successive governments to open up local authority work to outside competition. Under the Conservative government's Compulsory Competitive Tendering policy (CCT), much work was transferred to private contractors; the Labour government's Best Value regime has had the same effect. Recently the large scale voluntary transfers of local authority housing stock, removing their maintenance

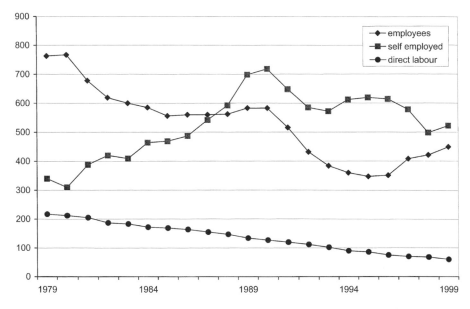

Fig. 4.2   Changes in employment by employment category 1979–1999 (thousands).

responsibility, can be expected to reduce the need for direct labour even further.

Secondly, Fig. 4.2 shows the remarkable change in the relationship between employed and self-employed operatives. In order to bring out that relationship more clearly, this chart incidentally excludes the category of workers defined as 'employees not on the register' for which estimates are made in the official statistics; it also excludes the administrative and professional groups.

From 1975 (and in fact before that) the number of self-employed increased steadily while the number of employees decreased; after 1990 numbers in both groups fell, with the overall decline in output; then after 1993 the self-employed figures rose again; only after 1996 do employees increase while self-employed decrease. The proportion of self employed shown in Fig. 4.3 also began to fall for the first time and although self-employed numbers rose again with increasing output, their proportion continued to fall.

These changes represent an important phenomenon but before trying to explain them it is necessary to define the terms used and to explore a little further the categories of employment in the industry. This is done in the next section.

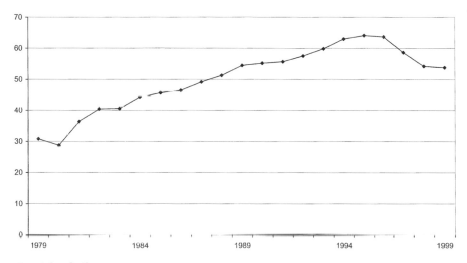

Fig. 4.3 Self employed as percentage of all operatives 1979–1999.

Modes of employment

The difference between employment and self-employment is an important one – but not a very clear one; self-employment itself is a category that has several variants. The following paragraphs attempt to elucidate some of the distinctions.

Employed workers

Employed workers are on the payroll of a main contractor or subcontractor company which is responsible for meeting all the legal obligations of an employer – including deduction of tax (PAYE) and national insurance contributions, ensuring healthy and safe working conditions, and providing holidays and sick pay. Employees are also better protected by legislation in areas of dismissal and redundancy. Apart from the requirements of the law, employers may also provide other benefits to their employees, either as matters of independently determined company policy or as the results of agreements with unions.

Self-employed workers

Self-employed workers are responsible for their own working conditions and for making their own tax and national insurance payments; they are not, in principle, eligible for paid holidays or sick pay and they do not automatically come under many other provisions and protection of employment legislation. However, within this general category of self-employed in the construction industry there are two further divisions and many possible different employment arrangements. The two subcategories are sometimes referred to as the 'bona fide', i.e. genuine, self-employed and the bogus (or even illegal) self-employed.

Genuine self-employed tradesmen

Genuine self-employed tradesmen approximately meet the accepted legal conditions for self-employment (though these are not entirely clear), the legal distinction is between a contract of service (an employee) and a contract for services (an independent contractor). In the construction industry the indications that a person is genuinely self-employed will include the facts that they:

● determine their own hours of work
● organise work on their own terms
● work on different jobs or for different contractors
● provide their own tools and equipment
● are paid for work done on completion of a job (i.e. not a wage)
● employ others or subcontract work.

So for example, a man owning a van and a range of tools, doing jobs for householders or other small (non-building) clients, paying his own income tax and possibly employing a secretary and a labourer is genuinely self-employed. He could of course work on occasion as a subcontractor to another contractor, he could even work as a labour-only subcontractor – but he would still be legally and for tax purposes genuinely self-employed (though his tax position would now depend on his turnover and certification under the Construction Industry Scheme discussed below).

Bogus self-employment

Bogus self-employment is where a worker is hired by a contractor as a labour-only subcontractor but on a more or less permanent basis. The main contractor will pay on an agreed basis, but will have no statutory responsibility for tax payments, for national insurance, for holiday entitlement or for sick pay. The worker himself will not meet any of the normal legal definitions of self-employment but is in fact being treated as self-employed. It is this type of self-employment that has caused so much debate and as will be argued later is at the root of many of the industry's problems. The unions in fact have argued that over 75% of self-employment is bogus in this sense.

Labour-only subcontracting

There are many possible arrangements by which workers are hired – they may be taken on through labour-only subcontracting firms, which hire workers out to the contractors, maybe on a self-employed basis; they may be hired through other sorts of employment agencies, which are the formal employers, or they may be recruited directly by contractors – again on a self-employed basis. All these cases can be referred to as 'labour-only sub-contracting' – which is not therefore the same as self-employment but very closely related to it.

The position can become very hazy – so much so that it is not clear even to the courts whether a person is self-employed, employed by an agency or employed by the contractor. In some cases unions such as UCATT have argued successfully that apparently self-employed workers are really employees and therefore have the same legal rights as employees; but in August 2000, Costain won its appeal against the decision of an industrial tribunal which had ruled that a man employed through an agency had been unfairly dismissed. The appeals tribunal decided that as the worker (who happened to be a UCATT safety representative!) worked through an agency he was deemed self-employed and therefore could not be 'dismissed' by Costain. It was, as UCATT said at the time, 'a worrying ruling'.

Self-employment – flexibility or problem?

Self-employment and labour-only subcontracting of this sort have been growing in the industry since at least the early 1960s and have throughout that time been a constant subject of controversy. In 1967, a parliamentary Committee of Enquiry (the Phelps Brown Committee) was established to investigate 'the engagement and use of labour in building and civil engi-neering, with particular reference to labour-only subcontracting'. It came to the conclusion that better regulation was needed, but that the practice should not be outlawed. But it did describe 'the self-employed form of labour-only subcontracting' as 'officially condemned by both sides of industry'.[8] The report, excellent though it was, satisfied no-one. In 1973 the UCATT-sponsored MP Eric Heffer introduced a bill in Parliament to have the practice outlawed; the bill was opposed from the political right on the grounds that it would interfere with the freedom of the contractors to decide how to organise their work and it was opposed from the left by some who argued it would interfere with the workers' freedom to determine the conditions under which they wanted to work.

In fact the practice continued and continued to expand as Figs 4.2 and 4.3 show. The basic reasons are that despite the long-term disadvantages, the immediate and short-term advantages to both sides always seem to predominate. These advantages and disadvantages have been recognised for many years; those discussed in the 1968 Phelps Brown Report are no different from the ones usually identified today. They are, very briefly:

- for the employer: saving in costs of employment
  reduced responsibilities under employment legislation
  flexibility in meeting changing workloads
- for the self-employed: freedom to choose when and for whom to work
  possibility of increasing income above agreed wage rates
  possibility of avoiding some tax and insurance obligations.

However although these are the underlying factors that encourage self-employment, many other changes in government policy have made self-employment more attractive. For example, the introduction of Selective Employment Tax in that it made direct employment more expensive; the Employment Protection Act of 1975 had, in the opinion of many employers at least, the perverse effect of encouraging a reduction in fully employed labour as it increased the employers' obligations and costs. In the 1980s, self-employment was positively encouraged; local authorities, for example, were forbidden to require contractors to use only directly employed labour.

However, the major influence on the split between self-employed and directly employed labour over the last thirty years has been the Inland Revenue and the income tax regimes it has developed specifically for construction. From 1970 up to 1999, the self-employed subcontractors were taxed in one of two ways. If they held a '714 certificate' they were entitled to be paid gross and were responsible for paying their own tax and national insurance contributions annually. If they held an 'SC60', the employer deducted income tax at source, but the 'employees' were still considered as self-employed. The system had been introduced originally to prevent tax avoidance and to regularise the self-employment situation, but in fact tax avoidance probably increased, as did self-employment. It was widely recognised that abuses were common.

In the early 1990s the Inland Revenue and the Contributions Agency became particularly concerned at the scale of 'bogus' self-employment. It was clear to the authorities that very large amounts of tax and insurance payments were not being made. By 1996 the Inland Revenue was applying intense pressure on the industry to eliminate this bogus self-employment, its auditors were instructed to examine a company's books in detail and to test whether so-called subcontractors were in reality employees of the firm. Firms found to have been abusing the system could be liable themselves for all the back tax due.

The effect of this pressure was, by some accounts, quite dramatic. Many firms decided to take former subcontractors onto their payrolls and according to one account, 200 000 people moved from subcontractor to direct employees representing about a third of all the self-employed. The official statistics do show a fall but not on that scale; between 1996 and 1998 the drop appeared to be about 100 000 (see Fig. 4.2) and in fact there is some evidence that after the initial scare, companies decided to stick to keeping their 714-certificate subcontractors.

However a completely new system was introduced in 1999 known as the *Construction Industry Scheme (CIS)*; the 714 and SC60 were abolished and three new levels of tax status were introduced, CIS 4, 5 and 6:

- The holder of a CIS 4 card is self-employed but has tax deducted by the contractor or client for the jobs on which he is engaged
- CIS 5 is for companies with a turnover of over £5 million, who can be paid gross
- The CIS 6 is a certificate that allows a small firm or individual to be paid gross if they have a turnover above a certain threshold £30 000 for an individual and £200 000 for a limited company.

The scheme and the method and speed of its introduction provoked much controversy and was subject after protests and negotiation to some amendments. It is still not clear that it is yet in its final form. The controversy was mainly over the details of the scheme and the extra amount of administrative work it was going to produce, but many argued that there was a fundamental flaw in the nature of the qualifications for the CIS 4 certification. This would prove to be such a burden, it was said, that it would simply drive subcontractors back in to the 'black market of cash payments' and reduce the numbers available for official subcontracting work. The problems of training and keeping skilled workers in the industry would be made worse.

At present it is not clear what the long-term impact of the new tax deduction system will be. Its purpose of course is to reduce tax avoidance, not reduce self-employment, but as we have seen the two are closely related. Meanwhile there may be other factors leading to a return to more direct employment – including the need to improve training which we look at in the next section.

Training and skills

The problem

Concerns about shortages of skilled operatives and the inadequacy of training have recurred throughout the last century; there have been many attempts by the industry itself and by governments to resolve the problems. Yet still in January 2001 the Construction Industry Training Board (CITB) reported that, 'according to current estimates there is a shortfall between the supply of qualified new recruits and demand from the industry. This would appear to be consistent with reports of skills shortages in the industry'.[9]

Photo (6)   The problem that won't go away – headlines from *Building* 1997, 1998, 1999 and 2000.

The problems have been persistent for some obvious reasons. When the industry has been in a period of recession, work has been hard to come by, neither employers nor unions have been in a financial position or had any real incentive to expand training programmes. Employers will be laying off the skilled men they already have; their profits, if any, will be low so unless they take a long term view – always difficult in construction – spending on training will seem an unaffordable luxury. The unions have been extremely anxious to restrict available skilled work to fully qualified tradesmen and in the nineteenth century and early years of the twentieth century actually tried hard to limit the number of apprenticeships to prevent an oversupply of trained men.

In times of boom consequently, the shortage of skilled workers rapidly becomes apparent; there is then much discussion of the need for better training, there is some action and it has usually been through government initiative – but it is still to no one's immediate advantage to finance training schemes and even taking a long-term view there is always the fear that the next recession may be just around the corner. Those employers, usually large firms that took their training obligations seriously, resented the fact that others who did not could simply poach their skilled people or that the newly skilled, once they obtained their qualifications, could go freelance as labour-only subcontractors at times of high demand. The feature of construction that we looked at in the previous section – the extent of self-employment – made the position much worse as in general self-employed workers are not involved in training programmes.

The old apprenticeships

The critical question, which arises every time skill shortages are discussed, is who should be responsible for training? In the days before the development of the modern industry and right back to mediaeval times, the answer was clear – it was the duty of the master tradesmen and the trade guilds described above. If a boy wished to become a joiner or mason, for example, there was only one way. He would be apprenticed to a master for seven years probably from the age of about fourteen; he could then become a 'journeyman' – which simply meant he was allowed to work independently for wages – and ultimately could become a master himself with his own workshops, journeymen and apprentices.

The guilds not only determined the rates of pay and conditions of employment, as we saw, but also controlled the conditions of apprenticeship as a way of establishing high standards of craftsmanship and the guilds were extremely jealous of standards.

The period of seven years was established very early on as being the minimum time it took to acquire the necessary skills. The London Masons' rules of the fourteenth century, for example, required that

'no master take no prentice for less term than seven year at the least because why such as be within less term may not perfectly come to his art.'

The language is ancient but the point is clear enough. Just what it meant to be an apprentice in those days can be gathered from the translation of an indenture of 1371 in Fig. 4.4 – where the prospect of having 'servitude' doubled to 14 years must have been a pretty powerful incentive to good behaviour!

The seven year apprenticeship was actually written into law from the middle of the sixteenth century (by the Statute of Artificers, 1563, men-

---

. . . . this indenture witnesseth that Nicholas son of John de Kyrghly shall serve well and faithfully in the manner of an apprentice John de Bradlay of York, bowyer, as his master, and with him shall dwell from the feast of St Peterad Vincula AD 1371 to the end of seven years next following fully to be completed; and the aforesaid Nicholas the precepts of the said master far and near shall willingly do, concealing his secrets and shall keep his counsel. He shall not do him damage to the sum of sixpence per annum or more, nor know of its being done without preventing it to the best of his power or warning his said master thereof forthwith; the goods of his said master he shall not waste nor lend them to anyone without his will or special precept; he shall not play at dice; he shall not be in the habit of frequenting taverns, gaming houses nor brothels; he shall not by any means commit adultery or fornication with the wife or daughter of his foresaid master under pain of doubling his aforesaid years of servitude; he shall not contract marriage with any woman nor marry her during the term of the foresaid seven years unless it were with the will and consent of the said master; from the service of the said master he shall not withdraw himself by days or nights during the abovesaid term . . .

Fig. 4.4   Extract from an indenture of apprenticeship to a bowyer (maker of bows) 1371.

Source: quoted by John Harvey *Mediaeval Craftsmen*, Batsford 1975.

tioned above) and remained the standard in most trades until well after the statute itself was repealed in 1813 – in fact well into the twentieth century.

The two world wars and the intervening serious depression resulted inevitably in more changes to the way training occurred in the industry; temporary schemes were accepted by the unions to overcome skill shortages during the First World War. But apprenticeship seems to have become rather unregulated and haphazard by then. Attempts were made at reforming and regulating the system through the National Joint Apprenticeship Scheme after the Second World War.

By the time of the Phelps Brown Report referred to above, the standard apprenticeship was four years (reduced from five in 1965) and on average about 10% of the total craft trade employees were apprentices. There were many others on informal apprenticeship-type training, but there was very little training indeed for non-craft operatives.

A training system without all the answers?

However, by that time the Construction Industry Training Board had been established (1964). It was hoped that at last a real solution had been found to the training problem; the Board could take overall responsibility for training to ensure a steady flow of skilled people into the industry. The scheme was to be financed partly by government subsidy but mainly through a compulsory levy on construction firms over a certain size (currently with turnovers of above £61 000).

The Board was not set up actually to run training schemes but to oversee, improve and bring some logic into the range of training systems that existed; though it did at an early stage set up its own national training college.

The Board's activities seem to have been controversial from the beginning. The compulsory levy was not popular. Firms objected to paying for the same reasons that they often were reluctant to undertake training themselves – they could not guarantee to see the benefits of the training themselves. Many argued that the CITB did not give value for money. There were proposals to abolish the Board altogether in 1993. It survived but there are still regular calls from disgruntled contractors for its major reform or abolition.

In fact the CITB seems to have been active in generating new initiatives, all attempting to improve the range of training routes available, and quickly moved to incorporate general developments in technical education and central government initiatives into the training programme.

There have been many new initiatives: expansion of off-site training (in the technical colleges and more recently through the Training and Enterprise Council's setting up of specialist training centres; expansion of training programmes; the development of construction National Vocational Qualifications and other certification schemes. It has expanded its training programmes to non-craft employees, supervisors, non-traditional types of work.

Yet the fundamental problem has not gone away. The number of trainees registered with the CITB fell dramatically from 135 000 in the mid 1960s to under 50 000 in 1988 then hovered around 30 000 during the 1990s, with a slight increase toward the end of the decade. In 1994 the Board published a report setting out a new training policy offering seven different routes to qualification for new entrants. The proposals were not fully implemented (partly because of a change in government training policies).

Today the board offers several routes and forms of training which lead to qualifications such as NVQ level 2 or above, most combining periods at college with periods of employment in the industry. An alternative way for new entrants is through contractors' own apprenticeship or training schemes such as those currently advertised by Carillion and other companies.

The New Deal programme introduced by the government to help unemployed people back into work (not just in construction) has also been used quite successfully by some companies; the industry was extremely sceptical about it at first, as it essentially involved taking young unemployed people who might have had no interest or motivation. The scheme has recently been revised and is being greeted with more enthusiasm.

As well as initial training programmes the CITB runs, supervises, subsidises or in other ways supports dozens of different schemes for widening and upgrading of skills. It has recently introduced what it sees as an important new initiative for training people already working in the industry, On-Site Assessment and Training, which had already signed up nearly three hundred companies in its first year.

Yet in spite of all this activity the criticisms and complaints continue and skills shortages are still with us; training, it is said by many including Sir John Egan, is still woefully inadequate. The CITB's own statistics show the extent of the problem with particular training shortfalls in specialist trades and civil engineering, confirming that it is still a real problem (see Table 4.1).

However it is not simply the shortfall in numbers, defined in standard trade categories that is the problem. The Egan report emphasises the need for new forms of training for multi-skilling and a need for much improved and more extensive training in supervisory skills. Detailed research done at the University of Westminster comparing British with European experience underlines the extent of these same inadequacies. Whereas the proportion

Table 4.1   Forecast of skills shortages 2001–2005

Annual average 2001–2005	Building trades	Specialist building trades	Civil engineering trades
New intake of trainees	28000	770	540
New output of trainees	21000	770	540
Required craftspeople	23000	5770	4900
Shortfall in formal training	−2000	−5000	−4360

Source: CITB

of British construction operatives classified as labourers is around 35%, it is only 5% in Denmark and 7% in The Netherlands; 'In these continental countries labourers have become a marginal group as increasingly only skilled workers are employed and the emphasis is on workers being able to plan and control their own workload.'[10] There seems to be some consensus on the root causes of the problem and slightly less on what should be done.

As mentioned before, fluctuations in construction activity are an obvious and well documented disincentive to development of training programmes in the industry itself; shortage of skills today is partly the result of the reduction in training which accompanied the downturn in activity in the early 1990s. Figure 4.5 shows an example of the changes in shortages of one

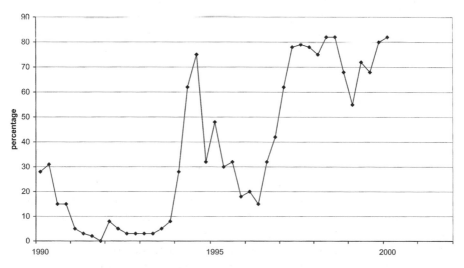

Fig. 4.5   Percentage of companies reporting shortages of bricklayers.

group: bricklayers. What is interesting about this graph is that there was apparently no shortage just after the peak of the construction boom in 1991/ 2. Construction output had at that time been growing steadily for nearly ten years (see Fig. 2.5, Chapter 2). It seems that people left the industry as demand declined and with the next rise in activity, the shortage became severe.

The second fundamental reason for inadequate training within the industry itself is the preponderance of small firms and even more sig-nificantly the high level of labour-only subcontracting discussed in the previous section. One idea behind the Construction Skill Certification Scheme launched in 1995 was that main contractors could demand that subcontractors used only workers with the appropriate level of certified skill, but small firms themselves often cannot afford to take on trainees, for fear of losing business to others who can operate more cheaply.

One of the solutions widely proposed is a greater use of prefabrication in order to reduce the range of skills required; the idea is dismissed by authors of the Westminster University report as quite fallacious as prefabrication simply means that different skills are required. Nevertheless there remain many in the industry who see this as part of the way forward.

More major contractors seem to be moving back to employing people on a full-time rather than subcontract basis; if this is a sign of real change and if many of the criticisms and proposals for reform are heeded we may begin to see real improvements.

Of course the best training schemes in the world are not going to increase the numbers of skilled people if the industry cannot attract them in at the beginning. This is another big issue which there is not space to explore here. Suffice it to say that there are now many initiatives by organisations and individual companies to make more effort at recruitment, and at improving the appalling image of the industry. In particular there are drives to attract more women and indeed a Women in Construction organisation and to bring in more people from ethnic minorities. The attempt to bring in more women seems particularly fraught with difficulty; the proportion of women, expressed as a percentage of the total industry workforce has hovered around 12% but as a percentage of operatives, it is very much lower – about 3%.

Underlying the problem of recruitment is the image of an industry which is dirty and dangerous with a 'macho' and racist culture. All these things cannot be discussed here in the detail they need but the issue of safety has become of major concern and is the subject of the last section of this chapter.

Health and safety

On 21st May 2000, three workmen were killed when a tower crane collapsed at a construction site in Canary Wharf; only a few weeks earlier three men had been killed on a site in Hull. Both accidents were dreadful reminders to the industry and to the world outside that construction has had a dismal safety record. Figures published by the Health and Safety Executive show construction has the highest rate of fatal and non-fatal accidents to *employees* of all major industrial groups.[11] In terms of accidents to all workers it is actually not quite as bad as agriculture where there is a very high rate of accidents to self-employed workers. At approximately 400 accidents per 100 000 employees its rate of non-fatal accidents is twice as high as in manufacturing. Table 4.2 shows the number of fatalities over the past few years; the estimates for 2000 sadly show a rise after recent falls

Table 4.2 Injuries to employees and self-employed in construction 1996–2000

	Fatal	Non-fatal
1996/97	90	13720
1997/98	80	14591
1999/2000	65	14232
1999/2000	78	14931

Source: Health and Safety Executive (1999/2000 figures provisional)

What is particularly depressing is that we are still seeing such high rates of injury and deaths after many years of increasingly tight regulation designed to reduce accidents and improve health standards, some of it specifically concerned with the construction industry. The current regulations are encoded primarily in one major act of Parliament and several sets of regulations set up under various clauses of the act:

● The Health and Safety at Work Act 1974 which laid the groundwork for subsequent legislation and regulatory instruments. It set up the Health and Safety Executive and imposed a very wide range of obligations on all employers in every industry;
● The Management of Health and Safety at Work Regulations and the Management of Equipment Regulations 1992;

- The Construction Design and Management Regulations 1994 (known now generally as CDM) – and their Approved Code of Practice;
- The Construction (Health, Safety and Welfare) (CHSW) Regulations 1996;
- The Lifting Operations and Lifting Equipment Regulations 1998.

The CDM and the CHSW regulations both included provisions implementing the EEC directive on construction of 1992 and all the sets of regulations above were essentially extensions of the original 1974 act.

The CDM regulations – the major instrument directed specifically at the construction industry – laid down new obligations principally on clients, designers and contractors. Clients have to ensure that a 'planning supervisor' is appointed. His or her obligations are:

- to inform the local health and safety executive of any proposals for construction work lasting over 30 days
- to coordinate health and safety aspects of design and project planning
- to ensure that a health and safety plan is prepared by the contractor which indicates, among many other things, where health and safety risks are likely to occur
- to prepare a health and safety file to be kept by the client and available for inspection.

Designers and contractors are responsible for seeing that the health and safety aspects of their work are considered in detail. For example the designers are required to 'examine methods by which the structure might be built and analyse the hazards and risks associated with these methods in the context of his design choices (Code of Practice para. 60 p. 19).

Contractors are required to develop the health and safety plan in a way which ensures that all Health and Safety at Work regulations will be met and compliance monitored throughout the project. In order to do this they have the power to require all subcontractors to do what is necessary to guarantee the safety conditions are met.

The 1996 CHSW regulations are probably clearer and simpler than the CDM but they go further in that they impose obligations on everybody including employees and self-employed workers. Many of the regulations are common sense written into law (e.g. regulation 8 '... do not throw any materials or objects down from a height if they could strike someone'). Others require actions that might not otherwise be taken – for example the requirement to inspect any working position at a height before work actually starts.

There are two obvious questions to ask:

- why after this steady flow of regulation and advice have things apparently not improved very much?
- what is to be done to ensure they do improve?

In part answer to the first question it has been pointed out that there has been a general increase in awareness of safety issues at all levels. The CDM regulations which at first were greeted with scepticism and the usual complaints about increasing bureaucracy and cost have generally been accepted and incorporated, at least at a superficial level, into normal procedures. It has also been argued that if we look at the numbers of deaths per thousand operatives, the figures do show a decline since the introduction of CDM. So one sort of answer to the first question is that in fact things have improved – a little.

Nevertheless an absolute increase in deaths and serious accidents, which has as we saw above actually occurred, is hardly a sign of great success. The simple answer to the question of why there has been so little advance is that the regulations are still not taken seriously enough right across the industry; it is possible to walk onto building sites anywhere and find breaches of the code.

As is usual in the construction industry (and probably in all other industries too) everybody seems to blame everybody else. It has been claimed that architects have not taken their CDM obligations seriously enough.[12] The unions have no doubt that it is the contractors who are failing their responsibilities: 'the basic causes of deaths' have not changed since the Blackspot Construction Report of 1988. This found that building workers were killed during simple routine work because of a lack of planning and that management were responsible for 70% of deaths.[13] Yet another report says management-initiated safety measures are undermined by workers' carelessness.

One factor, again, that seems generally accepted as an underlying cause of the high accident rate is the level of subcontracting and bogus self-employment. Self-employed and agency workers are far less likely to have gone through any safety training and are far less likely to be aware either of their own or of the contractor's responsibilities. For example a union representative found on one site that none of the self-employed had any idea what or who CDM or a Planning Supervisor was. So to the second major question – what is to be done?

It seems to be generally agreed that there is no need for further detailed regulation. What is needed is:

- first, even greater awareness and the dissemination of best practice
- second more comprehensive training
- and thirdly more effective enforcement.

Some companies have already taken major steps towards creating really safe environments and site practices though it is interesting that specific examples quoted seem to have been driven by the major clients involved – such as BAA and BP, at whose site at Sunbury, managed by Schal, rigorous safety standards are enforced.[14]

Secondly there are many training programmes already established and they are being intensified. The Major Contractors' Group and the trade unions have action plans to ensure that from 2001 all workers on their sites go through a 'site-specific' health and safety induction process before they start work. The programme will be carried out through regular consultation and workshops; all this is part of a campaign coordinated by the Health and Safety Executive and the (former) DETR.

Following from what was called the Safety Summit Meeting of February 2001, the Major Contractors' Group, which includes most of those at the top of Table 3.3 in Chapter 3, produced a 'Health and Safety Charter'. This is in effect an agreement to take a number of steps towards increased safety including having safety committees on every site and banning the victimisation of 'whistleblowers'. Union proposals for roving union safety representatives to make random visits to sites were rejected by the contractors, though unions are determined to introduce a system of this sort.

The real problem is to ensure the good intentions are actually realised and spread right throughout industry. It may be necessary before the whole issue is taken really seriously by all that there will have to be more rigorous inspection and enforcement. Unless evasions are identified even heavier punishments for negligence will not be a deterrent. There have already been some heavy fines (£1.2 million after the collapse of a tunnel at Heathrow imposed at the Old Bailey in 1999) but often punishments have seemed trivial. Many inside and outside the industry are of the view that until there is a crime of corporate manslaughter, and individuals are also found criminally negligent, the toll will continue. For the fact is that safety costs money and where competition is fierce and profit margins are slim, corners will be cut.

There is a related issue which deserves fuller discussion than is possible here, that of general health and well-being of workers on site. It seems to have been generally accepted that sites have to be dreadful places to work, dusty, dirty, muddy, cold in winter and badly serviced with toilets, washrooms and canteen facilities. Obviously there is no way that the construction of a multi-storey concrete framed building can offer something like the quiet, carpeted warm environment of a modern office (or at least the manager's office), but there is much that could be done. Egan quotes Tesco's introduction of 'visitor centres, on-site canteens, changing rooms and showers'.

Not every small firm can do this, but the industry that creates most of the built environment should really be able to improve its own.

Conclusion

Bringing some of the themes of this chapter together there seems to be one overriding factor which distinguishes employment in construction from employment in other industries – the very high level of self-employment and labour-only subcontracting. This we saw is a significant factor in the inadequacies of training, and in the industry's poor safety record. Whether this is about to change radically remains to be seen. If it does change, whether it can survive a serious recession is another big open question. But there are optimistic signs of changes in attitude, at least at the top levels, which have been to some degree sparked by the post-Egan rethinking.

For further study

On the early history of labour relations in the industry, P.W. Kingsford's *Building and Building Workers* (1973) and C.G. Powell's *An Economic History of the British Building Industry* are both interesting and readable. The classic history is R.W. Postgate's *The Builder's History* – published in 1923 by the National Federation of Building Trade Operatives. Chapter 4 of Linda Clarke's *Building Capitalism* (1992) gives a fascinating account of the situation in London at the end of the eighteenth and beginning of the nineteenth centuries.

The Construction Industry Training Board's website is www.citb.org.uk from which it is possible to download their reports such as *Construction Employment and Training Forecast 2001–2005*. The Education, Training and the Labour Market research group at the Westminster Business School (University of Westminster) is due to publish its final report on its comparative study of skills and training in several European countries. On safety, the Health and Safety Executive website (www.hse.gov.uk) is excellently clear. For example, linking to statistics – construction gives access to their latest very detailed report on accidents. UCATT, *Construction Safety – Building a New Culture* produced for the Construction Safety Summit in 2001 is a good summary of the issues and has references to further sources. Again, on all the issues a search of the various magazine indexes and archives such as www.building.co.uk can yield a great deal.

5

Professionals and Managers

- The construction professions – a confusion of consultants
- The architects
- The engineers
- The quantity surveyors
- The managers

Introduction: fragmented professions

There are many different groups of people working as professionals and managers in and around the construction industry – and there are many different definitions of what a professional or a manager is. Table 5.1 shows the average numbers of professionals employed in the different occupational categories in the later part of the 1990s. Each group plays a vital role somewhere in the complex process through which construction projects are planned, designed, built and maintained. However for the purposes of this chapter we will concentrate on the four major groups which have traditionally controlled the design and construction process – the architects, the engineers, the quantity surveyors and also the managers of construction firms, who are not included in the table. This is not to underestimate the role of many others involved directly and indirectly in construction, but simply to impose some limit on the extent of the discussion.

The much criticised 'fragmentation' of the industry is apparent in many areas, self-employment and subcontracting discussed in the last chapter are examples. But one of its clearest manifestations is in the division of responsibility between the different professions and between the professions and the contractors. This fragmentation is exacerbated by the fact that architects, surveyors and engineers usually operate from outside construction firms as independent consultants. In most manufacturing industry, functions such as product design, cost and financial control, and management of the production process itself are all carried out by the same company – though of course consultancy of many sorts is widely used and 'outsourcing' is becoming more usual. But in construction,

Table 5.1   Construction professionals employed (average) Spring 1995–Spring 1998

	In construction	Outside construction	Total	% in construction
Total professionals	90097	226753	316850	28
Civil engineers	34508	38489	72997	47
Architects	3916	36585	40501	10
Town planners	3086	10364	13450	23
Building surveyors	14367	47803	62170	23
Architectural technicians	2422	13614	16035	15
Civil engineering technicians	5057	5374	10431	48
Draughtspersons	5676	54965	60641	9
Building inspectors	2226	2424	4650	48
Quantity surveyors	18839	17135	35974	52

Source: CITB

architects, engineers and quantity surveyors involved in a project have most commonly been independent businesses or individuals. The disadvantages of this system as well as its advantages have long been recognised and the new forms of procurement we examine in Chapter 7 may bring about major change, but it is still the most common way of working.

One of the most unfortunate aspects of this division between the professions is the misunderstanding and even the hostility it has sometimes generated – hence again that other common criticism, 'the adversarial culture' of the industry. A good reason for knowing a little of the history behind all this is that it might help us to understand why the problems exist, for the basic causes lies back in the nineteenth century when each group worked hard to develop its identity and differentiate itself from the others – particularly those they saw as not quite in their class.

The next three sections below describe each group of professional consultants in turn, outlining some of the key historical developments that have determined the way they work today and describing the current position – the training required, the functions they perform and some of the challenges they are having to meet. The following section then looks at a few issues of management in the industry before we make a preliminary attempt in the final section to understand how the relationships between the groups are evolving and how a more integrated system of working might develop.

The architects

Origins

The fundamental function of the architect, the design of buildings, is perhaps more widely misunderstood – even within the construction industry – than that of most professions. Architectural design is often seen as a purely aesthetic concern, the creation of attractive built form or worse, attractive facades, concerned with appearance rather than structure. Form and aesthetics are certainly major preoccupations of every architect, but architecture has always wider concerns. Good architects attempt the difficult feat of designing buildings which function well, provide comfortable and enjoyable environments, are structurally sound, use materials effectively and are economical to build. Their training covers all those areas and more, though in many schools it is conceptual design that predominates, sometimes to the exclusion of other important skills.

At the end of their formal education and training – usually seven years – young architects are still only at the beginning of their learning path; and after many years' practice they can still get things wrong. They are frequently criticised; and of course as in every profession some architects are better than others, but it is still true that no other group in the industry is so thoroughly trained in the difficult process of designing buildings. Perhaps the greatest and most damaging misunderstanding is that some buildings are too small or simple to need professional designers; it is certainly possible and cheaper to do without architects – most buildings are not in fact architect designed – but there are few which might not have been improved with some high quality architectural contribution.

In Middle Eastern and western societies, at least, there always seem to have been specialists concerned with building design but the range of their functions has varied enormously. There are even records from ancient Egypt where, for example, the architect Senimut was not only 'architect of all the works' for Queen Hatsephut of the XVII dynasty but also 'Chief Guardian of the Queen's daughter and Governor of the Royal Palace'. Below Senimut in the court hierarchy, however, were other architects performing functions still recognisable as 'architecture' today – drawing outline plans and elevations, specifying materials and as 'overseer of works', managing the job on site.[1]

In classical Greece, that is around the fifth and fourth centuries BC, there was no clear distinction between architect, engineer and urban planner and many architects operated as what we would call building contractors. Although some historians believe that Greek architects did not 'design' in the sense of drawing out details beforehand, others believe that in fact they produced fairly detailed working drawings, and inscriptions show that they certainly laid down detailed materials specifications. Greek architecture remained a model of building form for centuries (it still does – you can buy plastic classical pilasters and pediments for your front door today).

The Romans of the first few centuries AD are usually thought of as greater engineers than architects – but the two functions were not seen as essentially different. Architectural training seems to have been well developed. There were apparently three main routes to a career as architect. One was a basic training in the 'liberal arts' followed by service with an established master, a second was through training in the army and a third was by working one's way up through the civil service grades. The picture of an architect's education painted by the Roman architectural historian Vitruvius is not entirely credible but it indicates the kind of qualities that might have been expected and were to be expected again in nineteenth century Britain:

'The architect should be equipped with knowledge of many branches of study and varied kinds of learning for it is by his judgement that all work done by the other arts is put to the test. This knowledge is the child of practice and theory. Let him be educated, skilful with the pencil, instructed in geometry, know much history, have followed the philosophers with attention, understand music, have some knowledge of medicine, know the opinions of the jurists and be acquainted with the astronomy and the theory of the heavens.'

After the decline of the Roman Empire, this notional ideal of broadly educated architect seems to have faded across Europe. The architect became essentially a master builder and the distinction between craftsman and architect disappeared. It reappeared probably first in Italy during the fifteenth and sixteenth centuries when people like Michelangelo built their magnificent churches and public buildings, as well as villas for the wealthy merchants, in Florence, Rome and other cities. But the combination of master builder with architect remained common and even as late as the 1770s in Britain 'a craftsman could easily combine his trade calling with the practice of architecture.'[2]

By then, however, some architects or architect surveyors working generally for the aristocracy, the church or the government had acquired a rather superior reputation. They had revived the idea of an architect as someone who thoroughly understood classical and Gothic precedents but who also knew about materials and the process of building. Planning and drawing techniques had become highly sophisticated and architects designed buildings before construction in great detail. Men like Sir Christopher Wren, designer of St Paul's Cathedral and dozens of much admired churches, Hawksmoor and Nash had themselves become rich through their work and created a class of professional which ambitious young people with a talent for design were anxious to follow. They were also, however, major builders, themselves employing building workers and taking control of the construction work.

Architects lower down the social hierarchy, however, were not always popular and were often viewed with suspicion and disdain as were surveyors and engineers – basically because anybody could claim to be one or the other or all three; it was the craftsmen and master builders who were more likely to have gone through rigorous apprenticeships and could offer real competence.

Developing a profession

It was partly to get rid of that kind of doubtful reputation that most of the groups in the industry set up clubs and societies at the end of the eighteenth century and in the early nineteenth century. In 1791 an exclusive Architects' Club was set up by four men who were also well-known builders as well as architects (James Wyatt, Henry Holland, George Dance and Samuel Cockerel); its membership was restricted to the very few at the top who had already won some recognition (for example as Royal Academicians). A more inclusive Surveyors Club, which included both architects and surveyors was set up in 1792; but this was also essentially a London-based dining club rather than a professional organisation. Other local societies followed but it was the British Institute of Architects (founded originally as the Architectural Society in 1831) that marked the real move towards the modern profession. The Institute received the designation 'Royal' from Queen Victoria in 1866.

Twenty years later it made a critical decision that was to affect the whole future of the way the construction process was managed. From 1887 no member of the RIBA could hold a profit-making position in the building industry. This had (at least) two unfortunate consequences: builders did not have their sons trained as architects as many had done before; secondly architects could not run construction firms, they could only be employees. A ditch had been dug between architecture and construction management. The situation did not change until 1981 but by then the separation of roles

was deeply embedded. Yet the decision had a good motivation behind it –
to ensure architects acted genuinely on behalf of the client and did not have
a financial interest in the profits of the contractor.

The RIBA existed both to establish a high status for architecture and to
protect the public by ensuring that the title 'architect' implied a high degree
of competence. For both these reasons it worked to develop a system of
examinations through which competence could be tested and it fought for a
system of registration. After a long struggle, statutory registration was
finally achieved in the 1930s first through an act in 1931 and then through
the Architects' Registration Act of 1938 which established ARCUK, the
Architects' Registration Council of the UK, and restricted the use of the title
architect to those who were formally registered. ARCUK has now been
replaced by the Architects' Registration Board, set up by the Architects' Act
of 1997.

The qualification system

Initially young architects trained as articled pupils and took examinations
externally – if they took them at all – but from 1894 when the first school of
architecture was founded at Liverpool it was possible to take a full-time
course before going into practice. More schools were established throughout the country but it was not until 1958 that a full-time architectural
education based on a minimum 2 A-level entry became almost the only
route into the profession.

Today most British architects go through a basic training that lasts a
minimum of seven years. The most common pattern is three years to a first
degree, one year out in practice, a further two years leading to an architectural diploma and finally a year in practice. All courses have to be
recognised by the RIBA and are regularly reassessed. For the student,
architectural training is a long hard road; even though academic entry
standards are high, many students take more than the minimum time to
complete and many drop by the wayside altogether.

The length of architectural course has often been criticised but has been
defended by the RIBA as essential to produce even minimally competent
designers. Courses are studio based, that is, students learn through the
actual process of producing designs of increasing sophistication. Courses
also involve formal teaching in architectural history and theory as well as
the technical subjects such as environmental science, structures, materials
and building technology. The old ideal, going back to Vitruvius, of the
widely educated *and* technically competent architect still survives.

After completion of their training, most young architects will find themselves working in small to medium sized practices where they will still have plenty to learn about the real problems of dealing with clients and contractors and of designing or contributing to the design of buildings which will actually be built (unlike those they will have been designing throughout their earlier years of training).

As the CITB figures in Table 5.1 show, there were 40 000 architects employed on average during the years 1995–1998; nearly 4000 were classified as working 'in construction', that is, employed by construction companies, with the rest in consultancies and private practice.

As in other construction sectors, architectural practices range widely in size, but even the biggest and best known are small compared with the giant contracting businesses described in Chapter 3. Many of these practices employ very few other professionals but there are also several thousand architects employed in multidisciplinary practices, sometimes architect-led but also in firms that are basically engineers or surveyors. Table 5.2 shows the numbers of architects working in firms with more than five employees, so it excludes the large number of very small and individual practices.

Table 5.2   Employment of architects by size and type of practice

Number of architects in one practice	Architectural practices		Multi-discipline practices	
	Number of practices	Total architects	Number of practices	Total architects
over 100	6	953	1	183
75–99	5	428	1	78
50–74	20	1126	2	110
25–49	21	723	1	27
5–24	30	492	20	205

Though fundamental, design has been only part of the architect's function, they have developed many other specialisms and offer many other services (the RIBA actually lists '103 things an architect does' on its website). Most important has been their role in the traditional building system (examined in the next two chapters) as essentially controllers of the construction process. It has been the architect's responsibility to see that the building is actually constructed in the way he or she intended – even though the formal contract for construction has been between the client and the contractor. It is this claim to be leader of the building team and manager of the construction process that has been increasingly questioned over the last thirty years or so and still causes controversy. The last section of the chapter looks at this issue again after describing the development of the other professions.

The engineers

Roman skills revived

Civil engineering as an activity is as old as architecture and, as indicated in the previous section, the two occupations were often not distinct from each other. The early engineers were concerned more with fortifications than their successors, but in ancient Rome, they were involved in designing and constructing grand civic buildings. It may be that the great engineers of the Roman roads, aqueducts, fortifications and amphitheatres were a different breed from the architects that Vitruvius had in mind in the quotation above but the distinction was never very clear. Whether called architects or engineers, their achievements were remarkable. The dome of the Pantheon, constructed in Rome in AD 120, had an internal span of 142 ft, a span that was not matched again for over 1000 years (by Florence Cathedral in 1367). The technology was well understood:

'As very dry and well compacted concrete (typical of Roman crafts-manship) was being poured on the dome's scaffold in horizontal layers from top to bottom, the builders introduced lighter aggregates, like pumice in the concrete of the upper part – and inserted empty clay amphorae which further reduced its weight. The Romans also under-stood the hoop action of the dome's parallels. In order to avoid the pouring of concrete on a horizontal scaffold at the crown of the dome, they left a circular hole or eye there; the rim of the opening was built of hard burned bricks, well cemented by excellent mortar, since it had to resist heavy compression in acting as the common keystone of all the meridional arches of the dome.'[3]

Eighteen centuries later a few minor cracks have apparently appeared!

Throughout the centuries that followed the decline of the Empire, the capacity to build large structures – fortifications, castles and churches – was widespread, but there was little scientific understanding of principles until the explorations of the artist engineers of Renaissance Italy. It was even later with the publication of a book by the great French engineer Belidor in 1729, 'La Science des Ingénieurs' before was there a serious foundation for that

Photos (7) & (8)   Architecture or Engineering: the Pantheon in Rome was a massive achievement by whatever name. 7. From an engraving by Hedley Fitton. 8. From a sketch by Mario Salvadori.

application of science to structure which perhaps distinguishes modern engineers from their predecessors. The word 'engineer' itself appears rarely, if at all in English before the sixteenth century and in fact the techniques seem to have developed more rapidly in France.

The French 'Corps des Ingénieurs du Génie Militaire' was established in the late 17th century to be followed by the 'Corps des Ingénieurs des Ponts et Chaussées' (Corps of Bridge and Road Engineers) and the 'Ecole des Ponts et Chaussées' in 1720. By 1760 General Belidor, the author of the treatise mentioned above, was France's Royal Inspector General of Engineering.

In Britain, John Smeaton, said to be the first person to have signed himself 'Civil Engineer' on an official document, had become a Fellow of the Royal Society in 1759 – a real recognition of distinction.

The Institution of Civil Engineers was founded in 1818 (interestingly, *before* the Institute of British Architects) though this was, at this stage, 'merely a study association, having no codes of conduct'. Its application for a royal charter was the occasion for the most famous – if now slightly dated – definition of the engineer's task by Thomas Tredgold:

'The art of directing the great sources of power in nature for the use and convenience of man . . . the most important object in civil engineering is to improve the means of production and of traffic . . . it is applied in the construction and management of roads, bridges, railroads, aqueducts, canals, river navigation, docks and storehouses . . . ports, harbours, breakwaters and lighthouses.'

Engineers were seen as a tough and often rough lot but with vital skills, as this Victorian description of the Liverpool dock engineer Jesse Hartley illustrates beautifully:

'On the eve of the large undertakings now contemplated (the date is 1825) the assistance of an engineer of large ideas and great practical skill was desirable and such a one was found in Jesse Hartley, whilom bridge master for the West Riding of Yorkshire, who was appointed engineer to the docks in 1824 and for 36 years guided with a despotic sway the construction of some of the mightiest works of the kind ever erected. Personally he was a man of large build and powerful frame, rough in manner and occasionally even rude, using expletives which the angel of mercy would not like to record, sometimes capricious and tyrannical but occasionally where he was attached, a firm and unswerving friend.'[4]

Towards qualification – and status

But their status grew as their importance was recognised: the canals, railways, the bridges and the docks which were all essential to underpin rapidly growing trade and industry could not have been built without them: '... they were no longer the admired but a rough race of men that they had been ...' Several engineers were knighted and the novelist, Trollope, writing in 1867, had only slight doubts about 'the suitability of engineering as a profession for the younger son of a well connected country squire'.

The number of specialisms within engineering increased with new techniques and materials. One of the latecomers, developing only in the twentieth century, was structural engineering. Its initial concern was the use of concrete as a building material and the need to better understand its properties (2000 years after the Romans!) but it extended its interests and expertise to the whole area of structural stability in buildings. The Institution of Structural Engineers was established as a separate branch (based on the Concrete Institute founded 1908) in 1922 and was granted a royal charter in 1934. The Chartered Institute of Building Services Engineers (CIBSE) was created only in 1976 but in fact was the amalgamation of two much older organisations, the Institution of Heating and Ventilating Engineers founded in 1897, and the Illuminating Engineering Society, founded in 1909.

The engineering institutions were slow to adopt a proper qualification structure. For many years they were reluctant to accept examination as a proper way to determine expertise, preferring evidence of practical experience:

> 'Even the queen of the branches of engineering, civil engineering, with the high prestige of its consultant status and its responsibility for the design of immense undertakings, remained convinced until nearly the end of the nineteenth century that learning on the job and in the office was a sufficient qualification, which could not or need not be tested with any basic principles of or fundamental body of necessary knowledge, which were perhaps held not to exist.'[5]

All that changed in the twentieth century, as greater scientific knowledge of principles became essential for practising and particularly designing engineers. Like the RIBA, the engineering institutions eventually established a system of qualifying examinations and training moved substantially into the colleges and universities. Moreover, after going their separate ways for a while, all the various branches established a common qualification structure. (See the Institution of Civil Engineers website for a detailed description of the alternative paths to fully qualified status.)

The academic standard required, particularly at the honours degree level, is high and the engineer's skills are held in respect by others in the construction industry. Yet they still seem to have a collective inferiority complex. A report in the 1970s said that 'engineers in Britain lack social standing' and in 1982 the President of the Society of Civil Engineering Technicians wrote, 'In the last 162 years British engineers have not achieved the standing or recognition comparable with the members of the medical or architectural professions'.[6]

Their practical importance in the industry is nevertheless immense; no moderately complex structure can be built without their input. Many of them are working as employees of contracting firms but the majority are with specialist consultancies ranging in size from the major companies such as W.S. Atkins with over 1000 staff down to one-man businesses. Table 5.3 gives an indication of their distribution across firms of different sizes.

Table 5.3 Employment of engineers in construction consultancies

Number of engineers in one firm	Number of firms	Total engineers employed
over 1000	3	4526
250–999	5	3338
100–249	24	3863
50–99	16	1183
10–49	39	823

Quantity surveying

The measurers

The early surveyors in Britain were concerned mainly with measuring and valuing land, usually working for or advising large estate owners on buying or selling land. Some of them became estate mangers, developing expertise in estate improvement and the generation of income from landed property. This type of surveying was very much in demand in the eighteenth century with the development of canals – requiring the purchase of large tracts of land from owners who drove hard bargains. It was even more in demand with the development of railways in the nineteenth century when the work of surveying often became merged with that of civil engineering and might be done by the same people.

Quantity surveying, as a separate specialism, had its origins, according to F.M. Thompson (author of the only detailed history of surveying) after the Great Fire of London with the massive demand for new construction in the city under the system of measure and value described more fully later (Chapter 6). Costs were determined, when a job was finished, by measurers who calculated the value of the materials used and the labour time required to build. The measurers were normally tradesmen but gradually measuring became a specialism on its own. It became especially important with the development of new methods of contracting (see Chapter 7) which required prices to be agreed between architect, client and builders – on the basis of plans – before a building was constructed.

Thompson considered the construction of the new Houses of Parliament in the 1830s as 'a great landmark in the history of ... quantity surveying'. It was one of the first buildings to have been approved on the basis of 'detailed drawing (by the architect, Barry) and a bill of quantities and full estimates based on them'. The bill and estimates were prepared by Henry Arthur Hunt who had been chosen by Barry as a well-known 'surveyor employed very much by architects and builders in making estimates'.

From that period on the main specialist functions of the quantity surveyor were established: to estimate, from the detailed architect's drawings and specifications, the quantity of materials and labour required to build each part of a building. The bill of quantities thus produced became used as an essential document against which contractors could tender and hence prepare a priced bid. Although quantity surveying came to be seen as fundamental in Britain, it did not develop separately elsewhere and there are good reasons to suggest that the methods of estimating it implied were never economically sound.[7]

The new professionals

Like the other developing specialists – architects and engineers – the surveyors saw the need to establish organisations to define and defend their role. Although a Land Surveyors Club was founded in 1834 and attempts made in the same year (defeated by the opposition of the architects) to form a Society of Architects & Surveyors, it was not until 1868 that the surveyors established a professional institution – the Institution of Surveyors. It acquired a charter in 1880 (which was really an official badge of status) and, after several changes in title, became the Royal Institute of Chartered Surveyors in 1946.

As with the architects and civil engineers, the Institution moved towards a system of examinations, as a route to professional status. Its relationship with the established educational organisations remained ambivalent for some time; the College of Estate Management, the first educational institute dedicated to surveying, developed as a private organisation. Cambridge University offered a degree in estate management shortly after the end of the First World War and London University also offered an external degree in the subject from 1919.

It was not until the 1960s however that the decision was made that surveying should become a profession for graduates in which full-time academic education at a university or technical college should be normal. These aspirations were fulfilled by the development of surveying departments in colleges, the polytechnics and the universities. Most surveyors have now therefore been through a full-time, part-time or sandwich degree course before moving into practice. Like architects, quantity surveyors are employed in private practices (see Table 5.4) but also, far more commonly than is the case with architects, they are employed in contracting firms.

Table 5.4   Employment of surveyors in consultancies

Number of surveyors in one firm	Number of firms	Total quantity surveyors	Total building surveyors
over 500	3	1154	311
250–499	9	2086	257
100–249	13	1095	239
50–99	15	795	178
25–49	34	618	394
5–24	62	335	287

In the one hundred and twenty or so years of its existence, the surveying profession has therefore raised its stature, its prestige, its membership and its influence. But the specific function and future of the quantity surveyor have recently been called into question. Even by the quantity surveyors themselves:

'The rate of progress (in information technology) makes it almost inevitable that software will automatically be doing much of the work now done by people, particularly measurement ... this side of the business is about to disappear and the QS will go with it if we do not change' (from reports on a QS 'think tank' in 1998).

However, the profession will no doubt survive by adapting and moving into new fields as it has already done. Some firms are developing a wider expertise and identifying for themselves new roles (or at least new names) as cost consultants, construction cost advisers or service providers. They are increasingly taking on the function of project or construction managers. Cost and economic advice will still be needed and that is where the real expertise of the modern quantity surveyor lies. Nevertheless it has to be recognised that other professions such as accountants may also be able to offer services such as advice and that in most other countries not influenced by the British system, quantity surveying as a distinct practice does not exist.

The management of construction

The idea of construction management as a 'profession' is relatively recent and it is still the case that although people can take the professional qua-lifying examinations of the Chartered Institute of Building, management is not and probably never can be restricted by law (as in the case of architecture and medicine for example) to people who have particular qualifications.

As the number of general building and contracting businesses grew in the nineteenth century, the men who controlled them came from many differ-ent backgrounds. Some were trained architects but as we saw above, after 1887 RIBA members could no longer be directors of contracting firms. Many others had been originally tradesmen, particularly joiners, carpenters and masons. Some were engineers, some were materials suppliers and some came from outside building altogether. They all learned through experi-ence, no claims were made then for a 'science of management' and management principles were only beginning to be codified.

However, a number of builders did get together just like the architects and engineers to form associations – usually to protect themselves against the growing trade union movement. The earliest organisation was probably the Builders' Society, formed in 1834 – the same year as the Institute of British Architects. Its aim was 'to promote a friendly feeling and the interchange of useful information among those who are engaged in general building in and near London'.[8] Their real objective was to establish some solidarity against clients who refused to incorporate arbitration clauses in contracts. Later in the century it was the Builders' Society which negotiated with the RIBA to create the first forms of agreed contractual document.

Other organisations such as the Central Association of Master Builders and in northern England, the Builders' and Contractors' Association, were more directly aimed at solidarity against union pressure. The first nation-wide organisation, the National Association of Master Builders, was formed after a strike in Manchester in 1877.

It was the Builders' Society that came to perform the role of a professional institution, evolving eventually into the Institute of Building and achieving chartered status as the CIOB in 1980. Like the other professional institutions, it now has a clearly specified qualification structure and again like them its educational base has shifted to a large extent into the colleges and universities. To grant exemption from the CIOB's qualifying examinations, courses have to be accredited and are regularly monitored. The CIOB's professional qualifications are available in five functional areas – Facilities Management, Commercial Management, Project Management, Construction Management and Production Design Management. There are now over fifty fully accredited degree courses in the UK, most of which give exemption in the construction management area, but a few specialise in Project and Design Management.

The CIOB has had considerable success in defining a professional basis for construction management and in persuading the industry that its qualifications are a true indication of basic competence. It has over thirty thousand members of whom something like a third are fully qualified. However, although the value of these qualifications is recognised, people can move into management positions through many different routes and from many different backgrounds; contracting and civil engineering firms were run largely by men who had learned their trade by experience, worked their way up as tradesmen or, as in the case of Laing and the McAlpines earlier this century, inherited family firms. Although many of them would have served trade apprenticeships and many from the family firms had been educated at university, few had any specific academic training in the process of construction management. Many in the industry felt (and many still do) that this was unnecessary.

In many small to medium sized businesses, there is still a strong family tradition where sons or, less often, daughters take over from fathers after a long period of learning the business – they could be seen almost as family apprentices and they may or may not have taken any of the CIOB qualifications. At the other end of the industry, the trend over the last few years has been for the large companies to recruit from the same pool of people as other industries – people with management training or qualifications or from other professions: lawyers, accountants or managers with experience outside construction.

This is still an area of considerable controversy. Sir Alfred McAlpine has written several times of his grandfather's belief that unless you had worked for a long time in 'welly boots' on muddy sites, you would never make a construction manager.

There are of course many different levels of management from the supervision of a small building site to the running of a major contracting organisation. Different skills and knowledge are required and people progress from the smaller to the greater responsibilities. Learning on the job is clearly vital but there still seems to be at all levels a reluctance to accept that formal training can improve performance. The Egan report identified poor training facilities at the supervisory level as a crucial problem and the most outright and scathing criticism was for management at higher levels. There is still clearly much to be done.

Cooperation or conflict?

From these outlines of the development of the major professional groups we can see how they steadily moved apart during the latter half of the nineteenth century and the early twentieth century. From a time in the late eighteenth century when a man could be at once engineer, architect, surveyor and builder, the professions evolved into separate groups with rigid boundaries between them. As each profession sought to strengthen its own identity and its claim to a special competence, forming separate institutions and creating its own examination structures, the divide between them deepened. They not only developed different skills, but quite different cultures – different languages even; reading the various professional journals today takes you into different worlds, though they are all essentially about the same business – construction.

The historian Marian Bowley described the outcome like this:

> '... the separation of design responsibility from building responsibility enabled builders to neglect their own education in design, or avoid the employment of designers, and the architect to neglect his own education in building practice ... Equally, the development of the system of independent quantity surveyors enabled designers to neglect proper study of estimating.'[9]

Meanwhile (see the quotation from Marston Fitch below) the engineers were out on their own.

The development of these distinctions was not entirely negative and may have been inevitable. A mixture of practical and social factors all worked in the same direction. First the level of specialism required became higher as construction became more complex. The old craft skills combined with a little dilettante knowledge of design were not adequate for the major building projects of the late nineteenth and twentieth centuries, but there was also undoubtedly a degree of snobbery and chauvinism involved. This becomes very clear from many comments made by representatives of one group about another. Each developing profession needed to establish its social credentials, to show it was respectable as well as competent, and if possible more respectable than the rest.

Interestingly, the same kind of self-promotion and denigration of the other groups occurred in other countries. The American historian Marston Fitch identified in the US 'the great schism' in the early nineteenth century as new forms of building imposed demands that traditional architects could not meet.

'To cope with these problems society was creating a new professional – the engineer. As the name implied, he was the lineal descendant of the "mechanic" of Jefferson's day, but with the theoretical training essential to the mastery of much more complex processes, machines and materials. As contrasted with the architect, the engineer's assignment from the industrialist was the simpler; to build quickly, cheaply and efficiently, and to hell with the looks. By and large the early engineer was not a cultured man, he was neither trained nor paid to explore the many fine shades of aesthetic problems over which the architects argued so learnedly. As a system of formal education in engineering began to evolve ... the neglect of the "artistic" aspect of the building began to solidify into ... contempt.'[10]

Exactly the same could be said about the situation in Britain during the same period and if the engineers were becoming contemptuous of the architects, the architects felt the same about the engineers. In 1862 an architect by the name of W.H. Leeds spoke condescendingly of

'... the new branch of art whose professors are called civil engineers ... we would not quarrel with these gentlemen, some of them possessing talents of the very highest nature, if they would be content with practising in their vocation. In their designs, even the best they have produced ... there are many violations of architectural propriety.'[11]

In other words, keep off our patch!

Attitudes to surveyors were even more dismissive even though only a few years earlier 'architects' and 'surveyors' were almost interchangeable terms. In the late 1850s there were suggestions that surveyors might join a common professional group with the architects but it was not to be. After suggesting that the Institute of British Architects might open its doors to surveyors, the *Building News* in 1860 pointed out that

'Although architects and engineers disagree upon all points of professional practice, they are united in treating surveyors as interlopers. Gentlemen who style themselves architects may be heard now and then lamenting over what they term the intrusion of builders to an artistic profession, and its consequent encumbrance with greedy, pushy and needy competitors who take the bread out of the mouths of some and lower the social standards of all.'[12]

If the engineers, architects and surveyors could agree on little else they would all accept without question that 'builders' were somewhere lower down the social hierarchy – despite the fact that by this time some of the more successful contractors had reached powerful political positions.

The world has moved on, class distinction is not what it was and we might expect today a more rational and comprehending set of attitudes and behaviour. With so many interdisciplinary practices such as those listed in the tables above, and with the constant pressure from government for an end to the 'adversarial culture', these old antagonisms could be expected to disappear. The situation is probably better than it was and much of the current discussion may exaggerate the problem. Yet these recent quotations (and there could be many more similar) seem to suggest otherwise:

First, from a report on subcontractors by Coventry University: 'Subcontractors had low opinions of quantity surveyors ... in particular young QSs tend to be arrogant and speak inappropriately to the subcontractors'.

Second, a quote from an article headed 'Down with Designers' by Colin Harding, the chairman of a medium-sized contractor and a regular contributor to *Building*, known for his strong opinions on architects (and others). His article expressed views commonly heard from contractors of all sizes but hotly contested by architects:

'very few architects have the construction management skills to justify their claim to be team leaders ... On about 75% of (our) contracts the management and/or quality of the product suffers ... as a direct result of designers failing to do their job properly. Typical lapses are:

- specifying inappropriate materials and systems
- distancing the supply side from the client and the design process
- no understanding of buildability
- lack of knowledge of basic construction techniques
- unwillingness to accept budget constraints

- inability to manage the design process ... design information is often late or inaccurate

... the standard of professionalism among architects is still falling. They need to be managed themselves – by construction managers.'[13]

Two months earlier the newly elected president of the RIBA, Marco Gold-shmied, had declared that high on his agenda would be:

'reclaiming a central role for architects in the design and construction process and ensuring that all Part 3 architects are qualified to do it.'

There does not appear to be much common ground there and not much change from the old adversarial attitudes. It is interesting however that one of the responses to Harding's article was by an architect admitting that these problems occur and urging architects to rethink the emphasis of their training. The sad thing is that this particular argument has been going on for at least 50 years.

The wrong response, though, is to dismiss the importance of good architectural design altogether; good design is desperately needed. There is plenty of it around and plenty of evidence that things can be done well. All the finest and most innovative buildings of the last century have required understanding and mutual appreciation of the skills of others. The engineer Paul Westbury, who has worked on the designs of such complex buildings as the Millennium Dome and the Arizona Cardinals Stadium in Phoenix has been quoted as saying about these kinds of construction: 'Understanding technology is crucial; working from day one as a team is crucial and the truly successful architects recognise it; that for me is construction, that's how it has to be.' Even at much humbler levels, cooperation between architects, other consultants and contractors is actually an everyday occurrence in the industry.

Yet there are still deep and unnecessary misunderstandings. Many routes have been tried to overcome them. Universities and colleges have recognised that the problem can start right at the beginning of training; it is surprising how quickly students identify themselves as architects, engineers or whatever almost before they have started their course. Attempts have been made to develop common courses, integrated projects, courses with common modules; they have met with varying success and although a report produced in the mid 1990s recommended that this was not the way forward, unless mutual understanding and respect are developed more effectively at the student stage, professional relationships are always going to be difficult. Shared postgraduate training is another route which is available – such as the Design Build Foundations Project Team Leadership

Programme developing in conjunction with Henley Management College. It does happen in other countries, though not universally. One example is the Ecole Superiore Nationale at Strasbourg, where all students begin with the same engineering basic course before moving either into architecture or engineering specialisms.

There are many developments in Britain that might – just might – lead to change. Ironically one of these may be the use of design and build contracts (discussed in more detail later in Chapter 7). These were intensely disliked by most architects because they claimed, often justifiably, that they usually resulted in banal and poor designs. Now that the problems are recognised more widely, clients and contractors are using architects more effectively and architects becoming more sympathetic (though there are still problems – for example of contractors using distinguished architects to gain planning permission, then changing designer and detailed design[14]). Again, the development of 'partnership' as a form of procurement (also discussed later) may have a similar effect. The strong advocacy of the 'one stop shop' by firms such as Amec and the consultants Arup is another sign that times are changing. The difficulty is always going to be ensuring that each professional area receives it proper influence and that design for example does not become totally subservient to the need for minimum cost construction.

As the discussion in this chapter has perhaps demonstrated, the problems are deep seated and are by no means yet resolved. Achieving the 'integrated team' that the Latham and Egan reports have called for still represents a continuing challenge to the industry.

For further study

There are countless histories of architecture and biographies of individual architects but fewer which actually discuss the development of the profession as such. Exceptions include F. Jenkins' *Architect and Patron* and Barrington Kaye's *The Development of the Architectural Profession in Britain*. The only major history of surveying is F.M. Thompson's *Chartered Surveyors – the Growth of a Profession*. The Institute of Civil Engineering's excellent website (www.ice.org.uk) has an interesting brief history with reference to the main texts, as does the Institution of Structural Engineers' site.

There is an excellent discussion of the relationship between engineer and architect in Alan Holgate's *The Art in Structural Design – An Introduction and Sourcebook*.

Current information on all professional groups is available on their websites: www.riba.org.uk; www.ciob.org.uk; www.rics.org.uk; etc. Again some are much more informative and easier to navigate than others but all are worth exploring.

6 The Growth of Contracting

- Contracting – a curious way of working?
- From 'measure and value' and 'in grosso' towards the modern contract
- General contracting becomes the norm – together with:
  - competitive tendering
  - bills of quantities
  - subcontracting
  - beginning of standard contracts

Introduction

> 'It is difficult to see how any system more wasteful of technical knowledge, intellectual ability and practical and organising experience could have been invented.'[1]

This was Marian Bowley again, on what has come to be called the 'traditional system' of procurement and contracting. Her view was that the separation of function between engineers, architects, builders and quantity surveyors, which we have discussed in the last chapter, had deeply damaging effects on the whole construction process. Instead of a symbiosis of all those skills to produce a system that would constantly generate innovation and efficiency, it produced a system that had conflict and conservatism built in. Many other academic critics and many of the reports mentioned in Chapter 1 right up to and including Egan have made the same types of criticism. If the system was indeed so flawed it is curious that it has lasted so long; possibly it never has been quite as bad as its critics have claimed or possibly it just became so deeply embedded that radical change was never really considered. The history of its growth helps us to understand, if not always to explain, some of its defining characteristics.

This chapter describes the development of the system from its beginnings up to the end of the twentieth century. Obviously such a wide sweep in one chapter is bound to give a very superficial account, but in fact the main features of the industry changed slowly and they can be identified in broad terms without entering too deeply into the detail. The next section describes the emergence of the contracting system out of the earlier practices of the eighteenth century and before. The following section examines the features of the system as it matured and the final section of the chapter looks at developments in the twentieth century.

The early years

'Measure and value'

From the Middle Ages until the end of the eighteenth century, the usual way of having building work done was to employ craftsmen directly. If the client did not want the trouble of overseeing the work himself, he would appoint an agent, usually an architect, surveyor or one of the master craftsmen themselves (there was as we saw in the last chapter little distinction between them in the early days) to organise the work and employ the other craftsmen and labourers needed.

This remained the normal way of building for centuries though there were different methods of employment and payment. The oldest system seems to have been direct employment at daily wage rates. The wage levels were fixed first by the craft guilds and later by local magistrates.

But from an early date, and no one seems to know exactly when, some work was let out under a system known as 'measure and value'. Work was valued after it was completed according to an agreed set of prices and wage rates (usually the same rates as used for daily wage payments). The measurers who did this work were the precursors of the later quantity surveyors. Sometimes – and there is evidence for this as early as the thirteenth century – master craftsmen would make a contract for a building or parts of a building and then employ other craftsmen and labourers to do the actual work, but payment was still usually on measure-and-value basis, not for a price agreed beforehand.

There was however a third way of doing things. At first it seems to have been quite unusual but became almost standard by the early nineteenth century, it was known as contracting to do work 'in grosso' usually translated as 'by the great', meaning for an agreed fixed sum. The architectural (building) historian John Harvey found examples from as early as the fifteenth century. One case he quotes is the erection of seven shops in London in 1408; Master Nicholas Waltham, a carpenter, contracted to do all the carpentry except tile pins for £55 and John Teffe, a mason, undertook to build the underground latrines, wall chimneys and foundations for £20.

These were what would be later called lump sum contracts but the important difference from later practice was that they involved *separate* agreements with the various tradesmen; and the difference from the usual method was that the price was agreed beforehand instead of the work being valued when completed. On the face of it this might seem like a very sensible development. The client knew exactly what the building work was going to cost and the builder could be expected only to agree a price that would cover his costs and give him a fair return for his work, so both sides should be satisfied. In fact from the very early days it was strongly opposed by many clients and craftsmen – it seemed to have been generally felt that work done under the *in grosso* formula was the least satisfactory for producing good quality or a good price.

Sir Christopher Wren used this system frequently. One example was in the building of the library of Trinity College, Cambridge where one contract was with a Robert Grumbold to pave the cloister and build four arches for £206 10s 6d including materials but though he used the system he did not seem to like it very much. In a letter to the Bishop of Oxford he explained that there were three ways of getting a job done, 'that is working by the Day, by Measure and By the Great'. The first could 'tell me when they are lazy; if by measure it tells every particular and what I am to provide' but when work was done in the *in grosso* formula, Wren could not be sure that the employer was getting a good bargain, because as the work was not measured, the lump sum figure was bound to contain an element of guesswork. On the other hand he could not be sure that it was fair on the tradesmen particularly if the work was unfamiliar; they then 'doe often injure themselves and when they begin to find it, they shuffle and sligh the worke to save themselves'.

Wren's work on St Paul's during the 1670s gives a good indication of how things were done at the time. As well as being responsible for the design he took personal responsibility for supervision of the work, with the help of two assistant surveyors. On his advice the first masonry contracts were signed with two master masons. They in turn employed sixty journeymen (qualified craftsmen) and over a hundred labourers. There were many other contracts to follow but they would always be with different sets of tradesmen for different jobs.

St Paul's was hardly a typical case, but the system of employing craftsmen separately by any of the three standard methods was used on humbler jobs as well and it carried on well into the nineteenth century. Sometimes the client would buy all the material and pay for labour only; this was the case in the building of an estate of houses for the engineers Boulton and Watt in Birmingham in 1810 and there were many other examples. In other cases the craftsmen would supply their own material, they would then agree a price for the whole job beforehand or work on a value and measure basis. Occasionally it was not clear on what basis they *had* agreed. In a court case found by the historian C.W. Chalkin from 1808, a baker claimed to have contracted with a 'builder' to build him a house for 300 guineas. The builder then claimed the right to be paid by measure and value (presumably because he realised he had priced too low) but lost his case.[2]

Early contracts

Although early examples of fixed price contracts were mostly with the separate tradesmen involved with a particular job, the contract with one single person to carry out a complete project (through subcontracts or employment of others) was not unknown and it was this which was eventually to lead to the modern contracting system. One of the earliest identified (though it is actually with two people working together) is quoted by John Harvey in his history of the mediaeval craftsmen. The gaol in York was apparently in a pretty bad state in 1377 – 'rather about to fall suddenly and slay the prisoners below.' The sheriff of York (and others) contracted with Robert de Nounon and Stephen de Barneby to demolish part of the old building and build a new wing

> 'with all the speed they may ... The said Robert and Stephen shall provide and find timber planking, thakborde, iron nails, lead, plaster, brick and workmen' and do the job for 'thirty nine pounds six shillings and eight pence'.

There were even agreed stage payments:

> 'first and in hand eleven pounds, and when half done fourteen pounds three shillings and fourpence and at the end of the whole of the aforesaid work fourteen pounds three shillings and fourpence'.[3]

By the end of the 1700s, the method of contracting with one person (or business) to undertake a project for an agreed price was becoming more common though still disliked both by tradesmen and employers. In 1780 Warwick gaol was rebuilt under this form of contract – but the work was unsatisfactory and this was put down to the nature of the contract itself so that when an extension was required in 1790 the justices resolved 'that no gaol was to be built by letting a contract to a single person in future'. Craftsmen too objected to the system as it undermined their independence and reduced them to the status of employees. As late as the 1830s the early trade unionists were still arguing about their craft rights such as that 'no new building should be erected by contract with one person'[4] – almost identical words to those of the Warwick justices.

In spite of all resistance, however, that was the way things were going to be.

The great transformation – general contracting takes over

It is not possible to date at all precisely when the change took place. The building historian Linda Clarke speaks of the 'virtual disappearance of the craftsman builder' in London in the last two decades of the eighteenth century and there is no doubt that by the early part of the nineteenth century, the new system of contracting with one person (or firm) for a fixed price was well established. It was to remain virtually unchanged for the next 150 years. The old ways of working lingered on; they still do of course especially at the domestic level where people directly employ individual tradesmen, but for larger projects the contracting system became the norm.

Yet it was continually and bitterly opposed by people on every side of the business, the clients, the architects, the tradesmen and even, at first, the contractors, particularly the small ones. The objections raised against the system are interesting as they are so familiar still and apart from the change in language some of the comments of the early eighteenth century could have come straight out of the Latham report. Architects were concerned with the problem of drawing up designs and specifications in such detail and with such certainty that no variations would be needed and the agreed price could be met. One of them put the problem in this way:

> 'In the ordinary course of work (by which he meant the old system of measure and value), while the work is in progress there is the opportunity of considering and laying out the detail of construction; but under a contract in gross, the architect cannot, unless he has the contractor's consent, make variations from what has been settled, or order what has been omitted in the specification, without an alteration in the sum agreed. The contractor, taking advantage of these circumstances, generally contrives to make the business advantageous to himself; the employer commonly pays more than the full value; and bad work is the usual result of a contract in the gross'.[5]

Although from the late 1780s there were regulations requiring government departments – then as now major clients for construction – to use the contract in gross, the men in charge of procurement generally ignored them. In the Office of Works the successive administrators were against contracting for a fixed sum as well as against contracting for the whole project so they carried on in the old ways – employing the separate trades.

The most bitter opposition, which continued right throughout the nineteenth century, was from the tradesmen and other building workers; in fact according to some labour historians opposition to contracting was one of the main incentives to the formation of trade unions. A placard issued by the Liverpool Painters announced that they had joined 'the General Union of the Artisans Employed in the Process of Building in order to put down that "baneful, unjust and ruinous system of monopolising the hard earned profits of another man's business called contracting"'.[6]

Despite the opposition and the reluctance of clients and architects, the system spread rapidly during the early years of the nineteenth century until it became the normal way of doing business. It developed in both public and private sectors; the London Custom House (1813–1817), the various stages of the new Houses of Parliament (1837 onwards) and Pentonville Prison (1839) as well as other prisons, and asylums were all built using fixed price contracts let to general contractors. In the private sector major buildings constructed under the system included Drury Lane Theatre (1811), Covent Garden Market (1828), the Athenaeum Club (1827–30) the Ashmolean Museum in Oxford in 1841, and many other lesser known buildings, as well as most churches. By the 1830s clients seem to have come to believe that the system was the best way to contain costs within estimates and to get value for money. A correspondent to the *Architectural Magazine* in 1834 wrote:

> 'I would recommend every person intending to build to get the work done by contract ... I would contract even for a dog kennel.'[7]

There were several elements in the establishment of the new system:

- replacement of 'measure and value' by lump sum contracts
- the appearance of general builders or contractors
- the use of competitive tendering
- the use of estimates based on bills of quantities
- the growth of subcontracting
- the beginning of standard contract documents.

The first has already been discussed, the others need a little more description and explanation.

The new builders and contractors

The term builder was hardly used before the last quarter of the eighteenth century, being a 'builder' was simply not a recognised occupation. It was clearly still new in 1763 when this was written:

'... of late years the capital Masters of carpentry have assumed the name of builders for this reason, because they make an estimate total expense of a House and contract for the execution of the whole for the amount of their estimates; for they take upon themselves the providing of all materials and employ their own Masons, Plumbers, Smiths, etc. whereas formerly it was the custom for gentlemen and merchants apply to several masters in each branch and employ them in executing their plans.'[8]

By the early years of the 1800s, builders of this sort were operating in most cities. Some had started working life as craftsmen – usually carpenters, masons or bricklayers; others were originally building materials manufacturers or merchants who often kept their materials business but expanded gradually into contracting. Others still were from outside the industry altogether – merchants who had built up their capital in other trades: ship owners, importers and exporters. As these firms developed they often tended to specialise in certain types of construction, some going into civil engineering, particularly the booming railway business, later in the century. Others became essentially developers, building housing to meet the rapidly growing demand in the cities.

Many became extremely wealthy and acquired high social standing. Thomas Grissell who with his partner Samuel Peto won the first contracts for the Palace of Westminster described below became an Associate of the Institution of Civil Engineers and later Sheriff of Surrey. Peto went into railway construction, operating in countries as far apart as Russia, Argentina and Australia, he was a member of Parliament for nearly twenty years. Unfortunately his business collapsed during one of the many financial crises of the time.

The best known of all were probably the Cubitt brothers, Thomas and William. Their separate careers give us a good indication of how the new, capitalist, stage of the industry was developing. At first the two worked together mainly as contractors. In 1827 they went their separate ways, Thomas to concentrate on development and William on contracting. In 1828 William Cubitt employed over 700 people and told a Select Committee of Parliament,

'I employ workmen in every department of building; and having large workshops and ample accommodation for the workmen, there is no difficulty in getting the very best workmen in London and the best superintendents.'[9]

Like other large contractors, he manufactured and assembled components for his buildings in his own workshops – closer perhaps to the sort of industry envisaged by the Egan report than most contractors today. William Cubitt became Member of Parliament for Andover and later, two years before his death in 1863, the Lord Mayor of London.

By this time the large contracting businesses had taken on essentially the same formula as today and as we saw in Chapter 2 many of the modern firms can trace their origins back to Victorian times. The name of Cubitt survived, as part of Holland, Hannen and Cubitt until it was absorbed into Tarmac in the 1980s.

Competitive tendering

Not all of the projects let under the new system to these major contractors were won through competitive tendering; often contracts were negotiated with firms known to the client or architect. Tenders were called for through general advertisements in the trade and local press; other times only a small group of 'respectable' builders were invited to tender. Contracts could be let to a single selected contractor without competition particularly if they had worked for the same client before. For example the firm of Grissell and Peto won the third contract for the Houses of Parliament and was directly appointed for the following four contracts – all at agreed fixed prices.

In fact in these early days of fixed price contracting, the whole idea of competition was controversial; many of the people who objected to the contracting system were really opposed to competitive tendering rather than to the idea of a single contract on an agreed sum. It was competition they believed which was bound to lead to lower standards as contractors anxious to get work would enter bids they could not possibly match in reality without reducing standards, using cheaper materials if they could get away with it and putting excessive pressure on the workmen. It was also feared that contractors would distort the system in all sorts of ways from collusion on bid prices to outright corruption. These were fears which were to prove justified over and over again.

In 1837 a contributor to the *Engineer and Architects Journal* wrote:

'After gaining the contract the whole object is to get the work done, no matter how, as cheaply as possible; thus inferior workmen and materials are often employed, notwithstanding the vigilance of the Clerk of Works, Surveyor or whoever might have superintendence of the work. I am fully aware of the seriousness of the charge I am making and the answer will be that no respectable builder will do such things ...'[10]

In fact the idea of respectability seems to have been seen as a significant protection against the worst consequences of competition. An architect for one of the government departments explained as early as 1813 that 'if works be of sufficient magnitude to make it worth while to excite competition, tenders are called for by public advertisement ... *if the parties are respectable,* the lowest offer is accepted'.

A series of government commissions on building procedures produced reports that were sometimes in favour of competitive tendering while others remained doubtful, but as the system became more common in the private sector, government departments came round to the view that it was the best way of obtaining value for money. By the middle of the last century, the practice of competitive tendering on the basis of design and priced specification had become the usual practice. The best protection against abuse was now seen to be the careful pricing of the specification and close supervision by the architect or other agents rather than reliance on a builder's 'respectability'.

The priced specification and bill of quantities

The growth of quantity surveying as a profession has already been described in the previous chapter. One of their major functions was the production of 'bills of quantities' which came to be seen as an essential element in the new contracting system. The essence of the system was that clients invited tenders on the basis of completed designs; in order for the builders to price their tender and for the client to know whether tenders were reasonable, it was felt that every element of the design had to be quantified (in terms of materials and labour required) and priced.

In fact the idea of bills of quantities goes back to before the general development of contracting. F.M. Thompson discovered a reference from Ireland dating back to 1750: 'a true bill of materials ... a full bill made by me, John Payne, Clerk and Surveyor of Quantities'. As Thompson points out, what appears to be a bill of quantities calculated from drawings before the start of building work was most unusual for this time.[11] Yet by the 1770s something close to the modern bill was coming into use. A contemporary writer Skaife outlined a 'method of preparing an accurate bill of quantities based on drawings of a specific building ... or if no drawings given, make a drawing from the idea propounded by the gentleman of what sort of structure he would choose' (which seems to imply a strange concept of accuracy).[12]

By 1828, when a Parliamentary committee investigated the Office of

Works and Public Buildings, the practice had clearly become well established:

> 'Both Smirke (the architect Robert Smirke) and Cubitt (the builder William Cubitt) agreed that the efficiency of contracts in gross was utterly dependent on an accurate and highly detailed specification from the architect of what was required in the building, and a careful calculation on the builder's behalf of the material implications of the specification. Smirke held that it was impossible to draw up a sufficiently comprehensive specification: "it would hardly be possible to avoid some ambiguity in parts of it, which the contractor could not fail to take advantage of." Cubitt seemed to doubt whether a skilled calculator of quantities was really necessary.'

Thompson goes on to say:

> 'Both reservations were quickly shown to be misplaced, the new contract made rapid headway and the new style of quantity surveyor was soon establishing himself as an indispensable link in the tendering process.'[13]

It could be argued that in fact Smirke and Cubitt had actually put their fingers on two basic weaknesses of the system which were to lead to constant criticisms, like those of Marian Bowley quoted at the beginning and the end of this chapter. Other countries did not find that the detailed bill was so essential.

However, Thompson was certainly right in saying that the bill did become fundamental to the system in Britain – particularly when as we will see later, it became part of the required contractual documentation.

Subcontracting

In the early days of contracting, one element of a builder's 'respectability' was his possession of sufficient capital and enough employees to carry a job through without subcontracting – for subcontracting was again often seen (as had contracting itself and competition) as a very dubious practice, yet it spread inexorably. In fact it probably grew out of a much older practice, the system sometimes known as 'blood for blood' and later 'cross contracting' where, for example, a joiner employed to build one house and a mason hired to build another would employ each other in their different trades.

Large-scale contractors such as William Cubitt, Henry Peto and George Myers prided themselves on the fact that they employed craftsmen in every trade and could therefore carry through a complete project under their own management control, but even they found it necessary to subcontract some work. Yet it always seemed to be done rather secretively, as it was so generally distrusted by clients and architects. Sir John Summerson claimed that even in the 1860s there was no subletting, 'a practice frowned upon by architects as leading to irresponsible cut-price workmanship'. In 1850 a glazier was sentenced to two months' imprisonment for intimidating his employers, Messrs Fox and Henderson, by threatening that unless they came to a decent agreement with their glaziers, an advertisement would appear in the London papers saying 'the building in Hyde park was being botched by a system of subcontracting; it will therefore be worthless and unsafe.'[14] But the advantages of subcontracting to the main contractors and ultimately to the 'small master tradesmen' were bound in the end to out-weigh the objections.

Some of these advantages, which continue to be relevant, were:

- First, flexibility. In an industry where work was so uncertain and when obtained, so varied, it was difficult for any contractor to guarantee permanent employment for large groups of tradesmen; they could and did take on and lay off workers, as they were required. But the tra-desmen themselves were more likely to find constant work by sub-contracting work from the several different main contractors; and they could also continue to do smaller jobs themselves independently.
- Second, risk and liability. An advantage for the main contractor was that subcontracting could push risk and liability down the line. In 1854 when workers accidentally caused an explosion by cutting through some gas pipes, it turned out that they were in fact subcontractors of sub-contractors to the main contractor; in the action brought by the woman injured in the explosion against the main contractor the judge held that it was the last contractor not the first who was liable for negligence.
- Third, specialisation. The development of new materials and new components required new skills, for example patent glazing, iron and steel frames and gas, and later electrical, lighting. Marian Bowley suggests that this was an important way through which innovation was brought into the industry.

However, the disadvantages and the problems that subcontracting brought with it were to cause continuing controversy. They are still with us and will be discussed in more detail in Chapter 7, but in brief they were:

- problems of control and responsibility
- problems of planning and sequencing
- problems of contractual liability

- problems of fair payment
- problems of employment conditions.

Nevertheless subcontracting became a fundamental part of the whole building process, though the problems associated with it did not disappear.

The nature of the contract

Of course one of the crucial elements in the whole contracting system was the nature of the contracts themselves. Even in the Middle Ages, as the 1377 example above from York gaol shows, contracts could be extremely detailed; they were usually drawn up separately for each project but there obviously developed basic structures and forms of wording. The idea of a contract is to make it absolutely clear what obligations each party has but no-one ever yet seems to have devised a contract – certainly not a building contract – which actually succeeded in eliminating all possibility of disputes over interpretation and performance.

As the contracting system developed, it was the architect acting as agent for the building owner who came to determine the conditions under which work was let. When the rules of the RIBA in the 1870s effectively barred architects from acting as builders, as described in Chapter 5, the main reason was to prevent any conflict of interest and to protect the client; the architects were seen as acting for and on behalf of the clients.

The builders felt the whole system to be rigged in favour of the architect and the client. One source of complaint from the builders was that contracts made no provision for variations in price if extra work was required or materials prices rose and yet as they pointed out, it was impossible for architects and surveyors to get everything specified exactly in advance, predicting every circumstance that would arise during construction. Architects tended to see any variations as the result of builders' failure to programme their work properly.

The Builders' Society had been set up in 1834, as we saw in the last chapter, partly as a group to oppose contracts without arbitration clauses but it was many years before the builders and the architects came to any sort of acceptable compromise. Other organisations such as the General Builders' Association and the London Builders' Society were all concerned with contract conditions.

In 1870 the RIBA's Professional Practice Committee and the London Builders' Society agreed on the terms of a document called the Heads of Conditions of Builders' Contracts; its aim was to establish the basic outline and principles of a standard building contract, which could be varied to suit particular circumstances – though parties were advised to take legal advice before making any changes. The contract was revised in 1880, then withdrawn and replaced by another in 1895 which was issued solely by the RIBA because it omitted the contractor's right to make a claim in respect of delay and alterations. Bills of quantities were introduced as part of the contract in 1902 as one of two alternative forms; new versions were issued in 1903 and 1909, with agreement now between the RIBA, the Institute of Builders (successor of the Builders' Society) and the National Federation of Building Trades Employers. This version remained in use until 1931 when the Joint Contracts Tribunal was set up.[15]

A case study: the Palace of Westminster

We have already referred several times to the construction of the Palace of Westminster, which was one of the major undertakings of the mid-Victorian period and on which a lot of information is available. As we saw in Chapter 1, it was a project which vastly exceeded its budget but it was also one in which many of the characteristics of modern contracting become clear for the first time. So it makes an interesting case study in procedures of the day. P.W. Kingsford gives a detailed and interesting account of the contracts themselves in his book *Building and Building Workers*[16] on which the following paragraphs are based.

The young architect Charles Barry, who had already made his name with buildings such as the Reform Club and many churches, won an open competition to design the new building, against 96 competitors. The government had decided that the building had to be built in the Gothic or Elizabethan style and Barry chose Gothic, though he would have preferred to build in the Italian classical style. Interestingly his design was chosen 'on the beauty and grandeur of its conception' and not much attention was paid to problems like ventilation and acoustics 'because not enough was known about these subjects'.

The project was carried out through a number of successive contracts 'awarded by the government department concerned as a result of competitive tender or recommendation by Barry'. The two first contracts – for the coffer dam and foundations – were let to Henry and John Lee, the third contract was put out for tender to eight contractors recommended by Barry and was won by the firm of Grissell and Peto who, as mentioned above, were then given the following four contracts without further competition.

In the first of these contracts the prices of different parts of the work were set by the builder (who had shown they were reasonable) and agreed by the architect. In the fourth contract, however, the prices were determined by the government department and they were set slightly lower than the other official prices which had been arrived at through other competitive work. Although Grissel and Peto agreed to this, they found they could make no profit and managed to negotiate a new set of prices.

Kingsford gives lengthy extracts from the contracts, which specified in detail how every aspect of the work was to be carried out and how much it should cost just as in a modern priced bill of quantities. The work of each trade was specified separately. An example from the Mason's Section gives a flavour of the whole contract

---

'The plinth of the whole wall, as shown by the drawings is to be of the best granite from the Fogging Tor Quarries, Devonshire, or from the Island of Guernsey of a fine grain similar to the best Aberdeen or from Aberdeen, and of an even colour throughout, being entirely free from large crystals of quartz of felspar, as well as from redness or stains.

Every stone is to be fine axed on the external face so as to present a fair and perfectly even surface at least equal to a sample now upon the ground which has been approved by the architect. The beds and joints are to be full square for their whole depths, particular care being taken to preserve the outer arrises, so that when the work is set it may be close and solid throughout, without any packing; and no joint is to exceed one eighth of an inch in thickness.'

(As this was the outside wall, you can still check to see if the job was done properly!)

So we see here a system already highly developed with practices well understood by the parties concerned and in essence it was the system that was to remain unchanged into the next century. Yet although it was becoming increasingly set in its ways, it never ceased to generate controversy and conflict.

The twentieth century developments

Dissatisfaction over methods of procurement and forms of contract continued and intensified after the First World War. The building market itself was changing, but the system was not.

One of the most important developments was the growth of local authorities as clients, particularly for the mass public housing programme which followed the housing acts of 1919 onwards. The difficulties that became apparent in attempting to achieve a rapid rate of construction, with high quality and low cost, led to a number of investigations and reports, some of which pointed to possible new ways of organising the whole process.

Other significant developments were those mentioned in the last chapter – the consolidation of the professions with the emergence of the structural engineers as a separate group, the protection of the title architect by the registration acts of 1931 and 1938, and formation of the Institute of Quantity Surveyors. If anything this consolidation represented a further deepening of the divisions which had been identified as one of the problems of the traditional system of procurement.

The one major advance – or what should have been a major advance – in all this was the establishment of the Joint Contracts Tribunal and the production in 1931 of a standard form of contract agreed between representatives of all the parties to a building project. It certainly was an advance in that there was at last an agreed form which was seen to be fundamentally on the right lines and which with inevitably frequent amendments has remained the basis of the vast majority of building contracts in Britain ever since. However, it also tended to strengthen the traditional system, with all its weaknesses and some builders and clients wanted changes – a different approach altogether. Some of the largest contractors, including Bovis, in fact insisted on their own forms, feeling that the standard forms were still biased in favour of the architects.

During and after the Second World War the pressures for change became stronger and a wider variety of approaches were adopted. The Simon Committee in 1944, and twenty years later the Banwell Committee, both recommended changes in the letting of contracts by public authorities – including an end to open tendering. Authorities began to experiment with new approaches. Bowley identifies four new ways as characteristic of the post-war period up to the 1960s:

- First there was selective tendering where only contractors known to have the adequate resources and a proven capacity to do work of the kind required were invited to tender
- Secondly there were negotiated contracts – bringing the contractor in at an earlier stage – used by local authorities when experimenting with new forms of building techniques
- Third, serial contracts where contractors having successfully completed one project were re-engaged on subsequent ones

● Fourth, the package deal as it was called, now more commonly referred to as design and build, was used particularly for the mass high-rise housing programmes in the 1960s.

Interestingly, none of these was new; all had been used before as, for example, the serial contracts to Grissell and Peto at Westminster; they have all reappeared, sometimes with new names, up to the present. However a proliferation of contract forms continued, all in attempts to overcome the recognised inadequacies of the traditional system – so that by the time Sir Michael Latham wrote his report he could identify over thirty different varieties.

In summary, although over a period of some two hundred years a system of building had developed and been maintained with very little fundamental change, it had been constantly criticised and constantly tinkered with. Yet still in the late 1990s the industry was found to be 'under achieving and in need of radical change' (Egan). The current issues are explored in the next chapter.

For further study

Marian Bowley's *The British Building Industry – Four Studies in Response and Resistance to Change* is still a classic. It is not easy reading but is worth delving into. C.G. Powell's *Economic History of the British Building Industry* is more readable but less detailed.

P.W. Kingsford's *Builders and Building Workers* covers the period from the early nineteenth century to the 1960s but is best on the Victorian period, with interesting accounts of the leading architects and contractors as well as building workers.

Akira Satoh's *Building in Britain – the Origins of a Modern Industry* examines specifically the transition of the industry from small-scale to large-scale production in the nineteenth century. There are a number of studies of particular builders, including Hermione Hobhouse's study of Thomas Cubitt.

---

7 Procurement, Contracts and the Way Ahead

- Introduction: some different ways of working
- Traditional procurement route
- Management contracting and contract management
- Design and build: some pros and cons
- Contracts and subcontracts
- Latham, procurement and contracts
- The Egan report: onwards to partnering and prime contracting?

Introduction

Although as mentioned at the end of the last chapter there were a multitude of different forms of contract by the time the Latham and Egan reports came to be written and many different ways of organising a construction project, it is possible to categorise the various methods into three or four major types, each with many variations. The whole set of arrangements for ordering and managing a project have come to be called the 'procurement routes' and were usually divide into the traditional and the non-traditional, with non-traditional taken to include design and build, various forms of management contracting and more recently, partnering and prime contracts.

Although the traditional route remained the most popular in terms of the number of contracts let, the non-traditional routes had become more common through the 1980s and 1990s, especially for larger schemes. One of the reasons for this – though it may also have been partly a consequence – was the shift in power relationships between clients, architects and contractors. Clients, at least the major clients discussed in Chapter 3, had become far more influential, but contractors too had strengthened their position vis-à-vis the architects. This can be seen quite clearly if one reads accounts of the industry's problems in the 1980s such as Michael Ball's, where one of the difficulties in achieving efficiency is seen as the subservient position of the contractor to the architect.[1] The growth of design and build and management contracting have almost certainly shifted the balance of power away from the architect (some would argue with unfortunate consequences for design).

There was a steady decline in the use of the traditional contract during the 1980s from 78% to 58% of all contracts let; by 1995 this had declined to 43.7% and by 1998 to 28.4%. The use of the design and build method of contracting increased dramatically, going from a 10% share during the 1980s up to 30% in 1995 and to a 41% share in 1998. Management contracting had about a 10% share in 1998, with construction management a large part of the remaining 20%. Over 80% of contracts were based on a standard JCT form.

Before examining the impact of the Latham and Egan reports and other relatively recent developments, this chapter describes in outline these major procurement methods in their modern forms and describes briefly what have been seen as their advantages and disadvantages for clients, for contractors and subcontractors and for the efficiency to the building process as a whole.

Photo (9)   Constructing the team – every large project today is inevitably team effort.

There are two major stages to any building or construction project: it has to be designed and it has to be built. The differences between the methods of procurement are essentially:

- differences first in the way people are selected to carry out these functions; basically they may be appointed after some form of competitive tendering or in a non-competitive process
- differences in the sequence of events from preliminary design ideas to construction on site
- differences in the formal and the informal relationships between contractors, consultants and clients. The formal relationships are usually set out in a legal contract, the informal relationships are much more difficult to define but are just as crucial to a successful project.

There are obviously a large number of possible combinations and variants of these different ways of operating and yet the division into two principal groups, traditional and non traditional, is still a useful basic categorisation.

The traditional system

What is now called the traditional procurement route is more or less the same as the system described in the last chapter, which had fully evolved by the beginning of the twentieth century. But there had been some evolution and the system as it operated (and basically still operates) was as follows.

The first step is taking the decision that a new building is needed, approximately what functions that building will fulfil and what facilities are required; clients even at this early stage might have a fairly detailed idea of the kind of building they want in terms of layout, position, size and even style. The next stage would be to appoint an architect. The RIBA has for years provided advice to relatively inexperienced clients on how to go about choosing a suitable practice for the job; the architect might be appointed on the basis of recommendation or a number of practices might be invited to submit outline proposals and will be interviewed for appointment. For larger scale projects, an architectural competition might be held.

The architect will be appointed usually under the standard RIBA conditions of engagement – normally on the basis of a fee related to the value of the project – a system which has been criticised for its obvious tendency to encourage designers to inflate values. With the generally increasing spread of the idea that competition was the only route to efficiency, 'fee bidding' became a common practice, that is, architects cutting fees in order to get the job. Fees on traditional contracts ranged in 2001 from 5.5% on small contracts down to 3.7% on contracts over £20 million.

The project brief (that is the description of the function and expected components) will be developed, ideally with the architect and the client working together. Initial feasibility studies are made, so that the client knows whether the objectives are likely to be achievable within the budget.

The project will then follow through a number of stages, which have been outlined in the RIBA's 'plan of work'. Essentially these are, after the briefing and feasibility stages:

- outline proposals: e.g. general layout, outline design and type of construction. QWS, engineers, etc. might be involved at this stage
- scheme design – detailed planning of building, constructional methods, outline specification and cost
- detail design – every part of building designed and specified
- production information – including final production drawings
- bill of quantities and tender documents prepared.

Tenders will now be invited by open advertisement or from a list of contractors preselected on some predetermined criteria such as a good track record in the particular type of work. In general the lowest tender will be accepted subject to checking by the quantity surveyor that there have been no errors in the contractor's pricing – for example that all items have been included.

An important point to note is that at this stage, in the most traditional system, *there has still been no involvement by the contractor* because until the tenders are received and a decision to appoint is made the contractor will not have been identified. Once the contractor is appointed project planning and actual construction can begin; the contract is between the client (not the architect) and the contractor. During construction the architects or their agents such as the clerk of works will be responsible for monitoring progress, for giving any further instructions, for certifying stage completion (on the advice of the quantity surveyor) and for determining the final handover to the client.

This is in bare outline the basic traditional, designer-led system – but there are many possible variations.

First, the tendering process can vary. The project may be openly advertised and all tenders considered or only selected contractors may be invited to tender, or there may be negotiations with only one contractor – usually one who has already worked for the client on a previous successful project. Tenders may be invited in two stages, first when only outline plans are available. Contractors are invited to tender on the basis of major cost items and rates. Once the contractor is selected he can become involved with the design team and final costs and specifications can be negotiated; although this is a way of overcoming one of the main objections to the traditional system, the separation of design and building, it is not often used in this form.

Other variants within the traditional system are to do with documentation, contracts and the selection of subcontractors. Tender documents may be drawn up on the basis of specifications – without a bill of quantities or with only 'approximate' quantities and costs detailed. In this country such an approach is often seen as creating difficulties and too much risk for the contractor as he has no certain basis on which to tender, but it is common practice in most other countries.

There are a number of standard contracts available for use, the most commonly used is still the JCT version, the latest at the time of writing being JCT 98. But others are in use and clients can even develop their own. This issue is discussed later in the chapter.

Subcontractors may be chosen and employed by the contractor or they may be nominated by the architect. As explained in the last chapter nomination has usually been unpopular with contractors because it further muddles the lines of responsibility and can undermine the main contractor's control of the construction programme. It is in much less frequent use than it was.

The claimed advantages and disadvantages of the traditional system have been widely discussed and we have looked at many of the criticisms in the last chapter, but it may be useful to summarise very briefly some of the generally accepted pros and cons:

Advantages

- It allows time and freedom for architects to develop designs fully, in consultation with clients, quantity surveyors, engineers and others.
- Tendering on the basis of drawings and bills of quantities provides a common basis for competition making unrealistic bids (high or low) less likely. It can achieve benefits of competition (lower prices) but, through the documentation provided and the rules of tendering adopted, can in theory at least avoid some of its damaging effects (such as low quality) as it is clear to all parties exactly what is being tendered for.
- The design team retains control of the way the design intentions are realised. Although it is the contractor who manages the actual construction process, his activities are subject to instructions from the client (the 'employer' in the contract), normally acting through the architect. It is the architect, for example, who frequently provides or appoints the clerk of works and it is the members of this design team who hold

regular site meetings with the contractor to discuss progress and if necessary issue instructions.
- There are agreed methods of introducing and pricing variations in the design if necessary.
- (Perhaps one of the most important) The system is well established and well understood by architects, contractors, subcontractors and even clients.

Disadvantages

- The contractor has no input into the design or initial estimating procedure; the contractor's knowledge of his own methods of working, the skills available to him and the current market conditions for materials put him in a better position than the architect to judge some aspects of the buildability of a design but this valuable knowledge cannot actually inform the design unless the contractor is brought in at an early stage.
- As the design is virtually completed before the contractor is chosen, there is a long time delay between initial proposals and the commencement of work on site, leading to increased costs.
- The tendering process is complex and expensive and where too many contractors are allowed to bid the total costs involved are very high indeed. The cost of preparation of tenders by all these contractors has to be met somewhere; ultimately the costs of the unsuccessful bidders will be covered in the costs of any successful tenders made for other projects.
- The system, it is said, perpetuates the 'adversarial culture' of the industry – it leads to inevitable misunderstanding and conflict between architects, QSs, contractors and subcontractors, again leading to losses in time and increased cost.

It is worth emphasising that although the system has been subject to so much criticism it has still survived and has produced – in fact it is still producing – many fine buildings at reasonable cost and on time. It is not perhaps irretrievably hopeless and may still be in need of reform rather than rejection.

Management contracting and contract management

There are two methods of procurement which are usually treated as 'non-traditional' but which are really developments of the traditional system in that they are designer led and most of the design process is separate from the construction work itself. These are management contracting and contract management – confusingly similar descriptions of confusingly similar procedures. They are sometimes referred to as fast track methods because by allowing the design and construction stages to overlap, actual building work can be started earlier than under the pure traditional system. In both cases a 'manager' – usually a contracting firm – is appointed to look after the whole construction process for a fee and is in effect taking over part of the architect's traditional role; indeed the management contractor can be seen as part of the design team; an architectural practice can in fact be the construction manager. The construction work itself is split into 'packages' (such as foundations, brickwork, mechanical services, etc.) which are let to specialist subcontractors. It is possible for the early packages to be let before detailed design of later stages has been completed; this not only shortens the total project time, but allows for interaction between the manager, subcontractors and designers.

The main difference between management contracting and its derivative, contract management, is that under the first system the subcontractors' contracts are with the manager while under the second they are with the client – leaving the manager with virtually no risk. Contract management has been used mainly by the large organisations such as the big retail chains which have major construction programmes and considerable experience. However, it has remained a very small proportion of total contracts.

Design and build

The third major group of procurement methods in use during the latter half of the twentieth century was the various forms of package deal, that is, where a contractor took responsibility for both design and management. As we saw in the last chapter the package deal has a long history and was not unknown in the early 1800s. The most common version today is Design and Build which has since 1981 had its own special form of contract available – the JCT Standard Form with Contractor's Design. Under this method the client goes out to tender on the basis of a statement of requirements, that is, before any actual design has taken place. Contractors submit tenders which include outline design, expected construction time and cost; the client or its professional advisers assess the proposals and select the contractor who will then be entirely responsible for the completion of the work.

The advantages claimed for design and build contracts are basically that they overcome the major perceived disadvantage of the traditional systems; design and construction are brought together. This means that:

- the contractor has sole responsibility – the architect cannot blame the contractor and the contractor cannot blame the architect for problems and failures
- design and construction can overlap – leading to quicker construction and lower costs
- experienced design and build contractors may use standard systems which they understand well and are less likely to lead to constructional problems or cost overruns
- there is direct contact between client and contractor.

However, the growth of design and build created considerable controversy and antagonisms between the architectural profession as a whole and the contractors. The main reason of course was that design and build reduced the status and power of the architect, but there was also genuine fear, often justified, that, by devaluing the role of the architect, it also devalued the importance of design; as a result, it was argued, the system would inevitably produce poorly designed buildings. This was bound to happen if contractors did not use competent designers (perhaps believing, mistakenly that good design was either not difficult or did not matter); it was also bound to happen if architects, when employed, were not given sufficient authority.

The problem of quality was recognised by many clients and variants of basic 'design and build' were developed to overcome these difficulties. One was so-called 'novation'; the client employs an architect or other consultant to design the building up to a certain level of detail and then novates – transfers – the agreement with the consultant to the contractor; the contractor then takes responsibility for completing the project. Experience seems to be very mixed with neither architects nor contractors particularly satisfied with the results Then there are architect-led design and build schemes. There are schemes in which though the contractor in fact appoints an architectural team, it is one already familiar with the client's requirements.

In fact there is some evidence that the general quality of design and build schemes has improved and it is certainly the case that some excellent buildings have been developed under this system. This is partly a result of increasing recognition by clients and contractors that good quality design matters as well as recognition by architects that they have to get closely involved with the construction business. It is now possible (since 1981) for architects to be directors of construction firms. At least one contractor at the time of writing (David MacLean) is registered as an architectural practice. This combination of design and construction skills in one organisation should lead to high quality results but it obviously depends on how good the designers are and how much influence and control they have.

Contracts and subcontracts

Whatever the procurement routes used, a common source of concern to contractors, clients, architects and subcontractors has always been the nature of the contract itself. The last chapter described how over a very long period a more or less standard and agreed contract (or actually set of standard forms) was established under the Joint Contracts Tribunal. It is not necessary and would be extremely difficult to follow even part of all the developments, evolution and convolutions since then. There is now a massive body of case law surrounding the building contract built up on literally thousands of disputes.

The general trends, however, can be outlined without too much distortion of the extremely complex reality. Three trends are obvious:

- First there has been continual refinement of the standard forms in an attempt to match different changes and to meet problems that have become evident through frequent conflict and resort to the courts.
- Secondly it has become common practice for clients and contractors to make specific alteration in the standard contract to suit, so they believe, their own special circumstances and this has led to even more disputed situations.
- Third, even though the JCT form was accepted as a standard, other forms of contract were developed and used alongside. Some have been seen as clearer and more effective, such as the New Engineering Contract used for civil engineering schemes; the British Property Federation produced a contract in the 1980s and advocated a contractual system which was often praised but never very widely used.

The interim report of the Latham committee in 1993 identified 17 different versions of the JCT form in use plus another 19 non-standard forms of contract.

The idea of the contract of course is to set out the rights and obligations of the parties to it clearly so each party knows exactly what is expected, what are its rights and obligations. If contracts were perfect, there would be no disputes. But of course life – and particularly life in the construction industry – is not like that; in fact the more a contract clause attempts to define in detail every possible eventuality the more it seems to become open to various interpretations. Take the following relatively simple example, chosen more or less at random from the JCT standard form (1980 with Quantities); it is part of a sub clause defining the 'relevant events' which must be taken into account if a contractor is to be allowed an extension of time.

> 'The Contractor not having received in due time necessary instructions, drawings details or levels from the Architect for which he specifically applied in writing provided that such application was made on a date which, having regard to the completion date was neither unreasonably distant from nor unreasonably close to the date on which it was necessary for him to receive the same.'

It does not take much imagination to see what a field day this could provide for lawyers in the case of a dispute over time extensions reaching court.

In fact of course the contracts themselves recognise that there will be disputes. The JCT 1980 form made provision for the reference of a dispute to an arbitrator who has to be a properly qualified person and accepted as such by both parties. The decision of the arbitrator is binding on both parties but either party may appeal to the High Court on any question of law arising out of the dispute. Although the arbitrator system has worked reasonably well, many hundreds of disputes have still finished up, expensively, in court. The 1998 Construction Act introduced a system of adjudication (in addition to, not in replacement of, arbitration) which is referred to later on in this chapter.

When a number of contractors, clients and consultants were surveyed in the early nineties, there seemed a reasonably high level of agreement that standard forms were generally a good thing but it was felt that they tended nevertheless to create mistrust and encourage conflict. It was also felt that they were unclear and difficult to understand.

One particular area of bitter controversy was – and still is – the contractual relationships between the main contractor and subcontractors; there are standard forms of subcontract to be used with the JCT main contractor such as DOM 1 (short for 'domestic subcontract') and others for nominated subcontractors. However it is common practice for main contractors to use their own subcontract or amend the standard form with clauses which are often seen to disadvantage the subcontractor.

Examples quoted over the past few years include:

- clauses which give the main contractor the right to dismiss the subcontractor if in the contractor's opinion he is not progressing fast enough

- clauses making the subcontractor liable for the defects of any work carried out previously by others
- clauses making the subcontractors' materials the property of the main contractor when on site – but at the same time making the subcontractor responsible for loss or damage
- wide rights of set off – that is, allowing the main contractor to make deductions for what the subcontractor may think are unjustifiable reasons
- 'pay when paid' clauses which allow the main contractor to delay payment to a subcontractor until the main contractor has itself been paid by the client.

These are the clauses most strongly objected to and generally felt to be unacceptable.

Latham, procurement and contracts

In spite of all the attempts that had been made to mitigate the well recognised problems of the traditional methods of working, the situation at the beginning of the 1990s was still seen by clients, the industry itself and by the government as unacceptably cumbersome and expensive. There was the general feeling that the proliferation of procurement routes and forms of contract documents, and the apparently endemic 'adversarial culture' were all part of a critical weakness in the industry. As described in Chapter 1, Sir Michael Latham was commissioned jointly by the government and leading clients in 1993 to make a study of the issues. Though much of what the report had to say was overshadowed and to some extent superseded by the later Egan report, it had important things to say about those major elements of the procurement process identified as problematic above – such as the critical problem of the separation of design from construction, the selection of contractors, the nature of the contracts and the position of subcontractors. It is worth looking briefly at the relevant parts of Latham's analysis and recommendations.

The separation of design and construction

It is important to realise that Latham did not totally reject the traditional route but did reject the use of contracts in which design and construction are completely separated and where the contractor can have no input to the design process. This sounds like a contradiction but it seems to imply that the traditional route can work if all design work is coordinated including the engineering and services design from the start and some contractors' input is incorporated at as early a stage as possible. The report makes a strong plea for the use of the coordinate project information system (CPI) which relates design drawings to actual construction procedures. It was developed a long time ago but is not widely used, in Latham's view partly because it is still not taught properly, if at all, in schools of architecture.[2]

In fact, although the report seems to favour some form of project management or design and build system, as long as design standards can be maintained, it does not come down firmly in favour of one specific procurement route. Rather, it stresses that clients need to be aware of the implications, especially the risk implications of each procurement method; it could be argued though that his definitions of risk are limited. Design and build and package deals are shown as offering the least risk to clients – but this does not include the risk, for example, of finishing up with a poorly designed building.

Tendering and the selection of contractors

In the selection of contractors Latham was very clear as to the essential reforms required. The report quoted cases of absurdly high numbers of firms tendering for relatively small jobs such as 28 tenders received for a £200 000 installation of heating systems in a London borough's council flats; other reports were quoted as saying lists of over twenty tenders were not uncommon. The complaint quoted in Chapter 1 (p. 9) from 1847 was still being ignored 150 years later! The recommendations were (to oversimplify) that lists should be severely restricted – in the case of design and build to only three tenderers, with two reserves, and that selective tendering should be used throughout the public sector (provided the procedures comply with EU regulations on advertising).

The nature of the contract

Recognising that the whole contractual process had become far too complex, generated far too much dispute and aggravation and enormous legal costs which ultimately are a burden on the consumer (though providing a major income stream to the specialist legal profession), the Latham report suggested a way forward to a neater clearer system. The actual number of different forms was not seen as a problem in itself but they need to be more comprehensible and meet common criteria.

The report listed 13 ideal principles of a 'modern contract' and suggested that from the existing standard forms, the New Engineering Contract came nearest to meeting those principles. It was proposed that this should become the basis of a new complete family of standard documents which would cover all kinds of building work, including subcontracts. The report further recommended that the Joint Contracts Tribunal should be restructured and a new body be created to be called the National Construction Contracts Council.

Latham on subcontractors – the 1998 Construction Act

Many of the Latham report's specific recommendations seem to have disappeared in the mists of time but his comments on the relationships between contractors and subcontractors have had important consequences – mainly through the 1998 Construction Act. Although disputes between main and subcontractors was only one of the many issues the review committee was asked to investigate it was the subject that 'generated the most heat and correspondence'. The report made a number of recommendations for increasing fairness between contractor and subcontractor, thus removing some of the chief causes of dispute. It also recommended reform of the method of resolving disputes when they arose.

Both these recommendations were incorporated, after much controversy, into the 1998 Housing Grants Construction and Regeneration Act (generally known as the Construction Act). Its construction clauses were said to be 'the most far reaching attempt ever made to rid the industry of adversarialism and contractual abuse'. Two major provisions were:

● The introduction of a new faster and cheaper system for the resolution of disputes. This was to be achieved by the requirement that contracts would have to include provision for the appointment of an *adjudicator* who was to make a 'final' decision within 28 days. ('Final' is in inverted commas because it has not worked out quite like that.)
● The requirement that contracts should ensure fair payments systems; the most significant element here was the banning of 'pay when paid' clauses in subcontracts.

The act was backed up by a Scheme for Construction Contracts, which was a set of rules and procedures for ensuring that its provisions were complied with.

However since the act was passed there have been continual rumblings of discontent among main contractors and subcontractors about the way it has been working. Disputes can still finish up eventually in court, even when the adjudicator has made an apparently final decision. Subcontractors have been particularly angry at the introduction by some main contractors of 'pay when certified' clauses; this gets round the ban on 'pay when paid' but, the subcontractors argue, has the same effect of unreasonably delaying payment.

The arguments continue and amendments to the act and the scheme are expected. The Construction Act has to an extent been overshadowed by the implications of the Egan report – in fact it could be argued that if Egan's proposals (discussed below) became universally accepted, much of the Construction Act would become redundant. This is because under an ideal long-term partnership between client, contractors and subcontractors, the question of unfair payment systems should not arise.

The idea of partnership was raised in the Latham report and it underlay many of the proposals for greater fairness, but it was specifically discussed in only one short paragraph in a report of over 120 pages. It was Sir John Egan's report which really put partnership, or partnering, at the centre of the debate on the industry's future.

Egan and the shift to partnering

The Egan perspective

As we saw in earlier chapters, Sir John Egan was not afraid to criticise the industry vigorously, and to make some radical proposals. Much of his report's discussion and recommendations were concerned with the actual production process and they will be considered in the next chapter, but it also argued that the improvements in production could not be brought about without fundamental reform of the whole procurement process which he saw as counterproductive, cumbersome and utterly outdated.

Egan's recommendations in this area can be summarised very simply and briefly, but they were so radical that they deserve quoting in bold letters! They were:

on contracts

...**an end to reliance on contracts**. If the relationship between constructor and employer is soundly based and the parties recognise their mutual interdependence, then **formal contract documents should gradually become obsolete**. It adds: 'The construction industry may find this revolutionary'.

on tendering

a reduced reliance on tendering; cutthroat price competition and inadequate profitability benefit no-one.

These were in a sense negative recommendations; the positive replacement of the old system was to be *partnering* and *performance measurement*:

> 'with quantitative performance targets and open book accounting, together with demanding arrangements for selecting partners ... value for money can be adequately demonstrated and properly audited ...'

His arguments for these ideas are succinctly set out in the report, which is quite short and very readable (much more so than Latham). They come down to the simple point that experience has shown that such procedures can work and do work; they have been shown to be extremely successful in the motor manufacturing industry but they have also been shown to work in construction, particularly by firms such as Sir John Egan's own – the British Airports Authority.

It is this concept of partnering that has become the main indicator of 'Egan compliance' and is set, if we believe all that is written about it, to revolutionise the industry's methods of operating. The definitions and ramifications of partnering are therefore examined in detail in the next section.

Partnering

By the end of the 1990s quite clearly the most favoured way forward, in terms of policy if not quite yet in practice, was 'partnering'. The ideas behind partnering were not entirely new. They were similar in many ways to the proposals in the Banwell report for serial or negotiated contracts where a relationship was established between a client and particular contractors over a series of projects. A report on partnering was published by the National Economic Development Council in 1991 and an American report on the idea '*In search of partnering excellence*' eventually attracted attention and became influential in the UK. As mentioned earlier, the Latham report touches on partnering almost in passing (paras 6.42–46) and made the rather weak recommendation that advice in the idea should be given to public authorities.

Egan's definition of partnering was '... organisations working together through agreeing mutual objectives, devising a way for resolving any disputes and committing themselves to continuous improvement, measuring progress and sharing the gains'. The potential gains were said to be huge, up to a 50% reduction in cost and an 80% reduction in time – figures based on claimed achievements on projects completed for some large clients.

This general definition (valid for and based on the practice in other industries) was elaborated in terms of the specific characteristics of construction, both by Egan and by academic and official documents such as the National Audit Office's *Modernising Construction*, published at the beginning of 2001.[3]

Basing its definition on some published and some specially commissioned reports, the National Audit Office identifies two major types of partnering: Project Partnering and Strategic Partnering:

- *Project partnering* is where the client and main contractor come to a series of agreements *after award of the contract*, which define precisely what and when things should be done, and which commit them both to mutually identifying improvements.
- *Strategic partnering* involves an association between client and contractor over a series of projects, where lessons can be learned successively and applied to achieve continuous improvements.

An essential part of both forms of partnering is the inclusion in the system of at least the most significant subcontractors – an arrangement referred to in most of the documents as 'partnering the supply chain'. Client, contractor and subcontractors all agree to accept targets, agree to measurements of success and aim for continuous improvement. Instead of subcontractors being selected on the basis of lowest bids, they would be selected on their proved ability to produce value for money, to agree to innovate, to accept open accounting and eventually to become incorporated in what would be seen as a coherent team. The idea of partnering the supply chain is discussed further in the next chapter.

To summarise, partnering on a project or strategic basis involves all of the following:

- shared commitment to shared goals, complete openness and fair dealing
- measurable targets for improving quality, delivery time and cost, with efficiency gains shared
- 'open book accounting' – clients know the contractor's real costs
- agreed system of consultation and shared decision taking
- involvement of subcontractors and suppliers in the partnership.

Even though Sir John Egan dismissed the protests that it was not possible to apply concepts such as partnering to the construction industry, he recognised the difficulties. 'Partnering is, however, far from being an easy option … there is already some evidence that it is more demanding than conventional tendering, requiring recognition of interdependence between clients and constructors.' Of course if the concept is to be applied fully it needs to go beyond the subcontractors to the materials and component suppliers.

There is still much debate on partnering and how far it can be effectively applied and there seem to be many different approaches in practice, for example, although the Egan report suggested that partnering could do away with contracts and operate on the basis of agreed partnering charter, in reality it seems unlikely that contracts will be abandoned lightly. In fact there is now a specific Partnering Contract, the Association of Consulting Architects' PPC2000 standard form for project partnering.

Some people argue however that a 'partnering contract' is a contradiction in terms, that the whole notion of partnership is too wide to be tied down in contractual clauses. The most likely development perhaps is that traditional contracts will be used as a sort of backstop to partnering charters.

There have now been many case studies published illustrating the techniques of partnering and the improvements in efficiency. For example the Audit Commission's *Modernising Construction* includes several examples. One study is of Anglia Water's shift from traditional procurement, which was claimed to reduce capital costs of projects by 20 per cent. One element of their new system was the dramatic reduction in the number of contractors and consultants with whom it worked – from two hundred and fifty to nine – six contractors, two design consultants and one cost consultant.[4]

Prime contracting

At the beginning of 1999 the Ministry of Defence announced that it was to use a new form of contracting on all its construction projects – amounting to over £1.5 billion a year. The system was called prime contracting and was to be piloted on two sports centres, one in Aldershot and the other at Wattisham, Suffolk. These projects were part of the Ministry's *Building Down Barriers* initiative which was to give a single contractor full responsibility for a project from the beginning and aimed at achieving major savings not just on capital but on a building's whole life costs. The contractor was expected to have a well-established relationship with suppliers operating to high standards and to 'integrate the supply chain into the design process'. If the contractor made more than the expected savings the difference would be shared between the Ministry and the contractor. The prime contract is in effect an extension of the design and build route adding tighter controls on the whole process, requiring high levels of performance to be achieved throughout the life of a building (or other facility).

A year later the Treasury announced that all public sector projects would be expected to consider prime contracting alongside design and build and PFI as preferred procurement routes though they could still use the traditional system if they could show it produced value for money. However when the first major prime contracts were drafted, contractors saw all sorts of problems and claimed that far from following Egan principles it imposed too much risk on the contractors and would lead to further conflicts. The contract expected a high quality of supply chain management – but did not require any kind of partnering agreement. Nevertheless in early 2001 it was still the intention of the government to extend the use of the prime contract and it was also announced that some of the big private clients, including Sainsbury's, were to adopt a version themselves – a way of adopting long-term relationships with contractors. It was not clear to outsiders at the time how this would differ from long-term relationships which had already been established by many major clients well before prime contracting came on the scene.

PFI and partnership

The private finance initiative was described in Chapter 2, where it was treated as a form of public sector financing of capital projects. However it also implies a different form of procurement and a new relationship between client and contractors. Under the private finance initiative the client – usually a public sector body – is acquiring a package of services, not just a building, and is paying for those services over a long period of time, not as a capital lump sum. So the contracts can obviously not be traditional though there is no reason why the early design stages could not follow the traditional route. In fact the schemes are more akin to design and build, being now defined as DBFO – Design, Build, Finance and Operate, etc.

PFI was introduced long before the Egan report – so what is its relationship to partnering? PFI schemes are now being described as one form of public–private partnership, but that just attaches a partnership label, it does not answer the question. A clue to the real answer is found in the National Audit Office report referred to above which suggests that PFI offers opportunities for the spread of partnering practices throughout public sector bodies. The relationship between the client and contractor under PFI is by definition a long-term one, so the client can be expected to require similar long-term agreements between the main contractor and the suppliers of all the services required.

However, there are certainly some concerns that everything could be so wrapped up between the client and a limited number of contractors and subcontractors, that all the benefits of competition could be lost and may not be completely substituted for by performance targets and performance agreements. There will be limited opportunity to test the market and to let in other (possibly more efficient) contractors. It is probably still too early to tell how effective the PFI schemes will turn out to be as true, non-adversarial partnerships.

For further study

This chapter has been deliberately sketchy as descriptions of modern procurement techniques are widely available. James Franks' *Building Procurement Systems – a Client's Guide* published by Harlow Longman is a detailed and authoritative survey. The RIBA publication *Which Contract?* (RIBA 1999) by Stanley Cox and Hugh Clamp is an excellently clear discussion of procurement routes *and* contracts.

There are many guides to the individual contracts such as those by David Chappell and Vincent Powell-Smith on JCT Intermediate, minor works and Design and Build; and Brian Eggleston's on the ICE contract conditions and on the New Engineering Contract.

The Audit Commission's *Modernising Construction* referred to in the chapter contains one of the best summaries of the principles, practice and potential of partnering; it includes a research report by Norman Fisher and Stuart Green and a number of interesting case studies. It can be easily reached and downloaded as a pdf file from www.nao.gov.uk. Other important sites to look up in this context are: the Construction Best Practice Programme www.cbpp.org.uk and the Movement for Innovation www.m4i.org.uk.

8 Building Production – Site or Factory?

- Introduction: an outdated production system?
- Buildings, cars and economies of scale
- Technology and innovation in the construction process
  - mechanisation and traditional methods
  - innovation in the final product
- Standardisation and prefabrication
  - in housebuilding: to industrialisation and back – again and again
  - and other forms of construction
- Coordinating the process: managing the supply chain
- Integrating the whole process: some solutions

Introduction

The previous two chapters examined the development and practice of different procurement methods and contractual relationships between client, designer and contractor. The various procurement routes determine how the construction process itself is managed; and it has generally been argued that some routes lead to a more efficient production process than others and that the traditional route is particularly prone to the creation of delays and excessive costs. However, there seems to be very little research evidence to support any general assumption that well-managed schemes under a modified traditional contract cannot be built as efficiently as under any other procurement system. Claims for big savings in time and cost by using one system rather than another often turn out not to be comparing like with like and very few comparisons incorporate design quality into the criteria used.

In fact the actual process of assembling a building can look very much the same whatever procurement method is used; differences depending more on the scale of the project involved. However, that process itself has been as widely criticised for being outdated and inefficient, incorporating little technical innovation and still too dependent on a poorly organised and poorly trained workforce.

This chapter concentrates on some important aspects of the way buildings and other projects are actually put together and examines some of these criticisms in the light of the industry's actual practice. In discussions of the industry's construction as opposed to its procurement methods, the same issues tend to come up again and again. Why do so many projects go wrong in one way or another? Why does the process so often fail to run smoothly? Why keep reinventing the wheel by designing different buildings for essentially the same purpose? Is the industry technologically backward? Why have some methods of building stayed the same for hundreds of years? Why can't production of buildings not be more like the production of cars, fast, efficient and yet varied with constant and rapid technical innovation?

To this last question, the Egan report gave a clear answer – it can be and it should be. The report's final words were:

> 'What we are proposing is a radical change in the way we build. We wish to see, within five years, the construction industry deliver its products to its customers in the same way as the best consumer led manufacturing and service industries. To achieve dramatic increases in efficiency and quality that are both possible and necessary we must rethink construction.'[1]

As the history summary in Chapter 1 showed, the Egan report was not the first to accuse the industry of gross inefficiency, it was only the latest in a series stretching back over a century. Nor was it the first to suggest that the industry should take motor manufacturing and other modern high tech nology industries as a model for change. Egan vehemently rejects the claim that construction is in some special way different from the rest of industry:

> 'We have repeatedly heard the claim that construction is different from manufacturing because every product is unique. We do not agree. Not only are many buildings such as houses essentially repeat projects which can be continually improved, but more importantly the process of construction is itself repeated in essentials from project to project.'[2]

This chapter looks first at questions relating to the uniqueness of construction. In what ways is construction different? Do the differences in the nature of its output justify the differences in its methods? We then look at three important aspects of its methods which might be further developed to move it in the direction the Egan report proposed:

● the use of modern and appropriate technology
● the adoption of standardisation and pre-fabrication

- the efficient co-ordination and control of all the inputs to the construction process.

Looking at each in turn, it is suggested that in the case of the first two, the industry has in fact moved further than it is often given credit for; often it seems to have retreated and we need to understand why.

In the third area, management and control of inputs (such as materials and the work of designers and subcontractors) the inadequacies of many procedures are really difficult to defend. But even here there are rational reasons for the way the system has developed and the way it still for the most part works. It is certainly in this area that the Egan report concentrates its fire and for which it makes the most radical proposals.

However in all three areas there is room for improvement. The chapter concludes with a discussion of some of the current proposals and the post Egan developments by which the industry is, in principle, now being driven forward.

A building is not a motor car – is it?

Leaving aside the trivial and obvious differences between buildings and cars are there in fact many fundamental *economic* reasons why construction cannot take place in the same way as the production of cars or computers or washing machines? There are at least four possible ones:

- The first and most obvious one is the fact that buildings are very heavy and not too easy to move (though, as odd examples have shown, not impossible). The point sounds trivial but it is very significant – almost the only other manufactured products as heavy as buildings and which *can* be moved are ships and planes, which have their own means of locomotion.
- Secondly, following on from the first point, buildings and other construction projects are almost the only manufactured objects which have to be made mainly on their ultimate site; by definition this means that hundreds of them cannot be made in the same place, like cars in a factory. (Objections to this point are considered later.)
- Third – a high proportion of buildings are unique, designed for a specific location and a specific purpose; they are not standard objects. The main exception and a very important one, is the house, and there are others which are discussed later.
- The fourth point is the one already discussed in Chapter 2: construction is subject to large and sometimes quite sudden shifts in demand. It is

true that other industries are also subject to varying demand but for most the changes are usually at the margins. If you are producing five million Kit-Kats a day, a fall in consumer demand still leaves you with a very high output. The individual construction firm on the other hand, might find it has difficulty in securing any contracts at all in one period and then, if it survives, find itself swamped with opportunities which it cannot take up. All industries suffer if there are really deep recessions but many can ride out milder recessions by producing for stock or closing down a part of the production process. This is much more difficult in construction.

The economic consequence of these characteristics of construction is that *economies of scale are difficult to achieve*. The car industry has grown right from the early days of the first mass-produced cars through using the advantages of economies of scale made possible by the steady sustained growth of demand. It took Ford ten years, from 1905 to 1915, to produce a million cars; by 1924 they were turning out two million cars a year. Yet costs fell, quality and sophistication increased and wages were high. This process has continued in car manufacture ever since.

Economies of scale arise from many factors: the early factories depended on the division of labour (each worker doing one task repeatedly), continuity of production and increasing mechanisation. As output grew, higher profits made possible more investment in new technologies that made even greater economies and more advanced products possible. This same process has taken place across the whole of manufacturing industry. But it has not happened in construction (with the partial exception of housing – again) and it seems unlikely that it could do. The single Honda factory in Swindon is expecting to produce 250 000 cars a year by 2002; that is more than the total annual number of houses of all types built by all the housebuilders in the country. Of course if, improbably, the housebuilders all merged into one giant outfit, maybe the same economies of scale would arise. The likelihood of that happening and its desirability are both questionable – though as we saw in Chapter 3, a very high level of concentration may now be conceivable.

However as the quotations above show, the Egan report rejects the idea that the differences between construction and manufacturing justify the industry's failure to adopt modern techniques. It is certainly true that although a total mass production system may not yet be conceivable in construction itself, some of the important generators of efficiency common in other industries *can* be adopted. In many respects they have in fact been adopted already and the real question is have they been developed and applied in the most effective way? The next sections look at three of the routes to greater efficiency mentioned earlier: use of leading edge appropriate technology by a highly skilled workforce, standardisation of components and product, and efficient integration of all inputs. In each case we look at what has happened, what the current situation is and what developments might be made.

Technology and innovation

Mechanising the traditional way of building

Building is essentially a process of preparing a site, bringing in materials and components, forming materials into elements such as frames, walls and roofs, assembling ready made components, installing services and then finishing ready for occupation. That could describe construction at any time over the last few thousand years. What has changed over that time is the proportion of components made away from the site and the importance of internal services and finishes, both of which have increased.

What is often said *not* to have changed very much is the process of construction itself – the actual assembling of all the elements into a building. It does seem a little odd that men still build walls by putting one row of baked clay blocks on top of another and binding the two with mortar. They did it like that in ancient Egypt over three thousand years ago. Even mechanical devices such as hoists and cranes and other aids to high building such as scaffolding are not new. They can be seen in pictures of building from the Middle Ages.

In fact, a small building site today can still look more like a scene from one of those old prints than part of some modern industrial process, but these comparisons can be exaggerated. There were no tower cranes hovering above the Parthenon or the dome of the Rome capitol or swinging blocks of stone into place on St Paul's cathedral (which makes them all the more remarkable achievements) and there were no dumper trucks or men in yellow hats.

In fact even the apparently oldest tasks such as building with bricks have only a superficial similarity to their ancient predecessors. It has to be remembered that the modern brick is a mass-produced, quality controlled product; the economies of scale which are difficult to achieve in construction itself are considerable in brick making. The modern sophisticated manufacturing technology which may not be evident on the building site is normal in the production of materials and modern components – glass, plastics, cables, roofing, steel as well as bricks. When the raw material is widely available (such as clay) and the production process efficient, the resulting product is relatively cheap, which is why a good bricklayer today can still produce walling with an economic efficiency to match apparently much more advanced methods.

Because this kind of development in materials and off-site fabrication has partly been ignored, the industry has been accused of remaining technologically backward compared with most others, which have been transformed totally by technical change. However, improvements in materials and materials manufacturing, use of more prefabrication and the application of mechanisation to site work have all been going on for a long time in construction. Already by the middle of the nineteenth century new techniques of pre forming materials were also being adopted by the larger contractors, which at that period often made more construction components 'in house'. The largest general builders, such as William Cubitt in London, Messrs Holmes in Liverpool and Pauling in Manchester, had extensive workshops with steam driven machinery for every kind of timber and stone working: planing machines, circular saws, moulding machines for wood and similar machinery for stone.[3] At the same time there was much experimentation in applying technology to work on site. For example, the steam engine was applied at a very early stage in its development to lifting heavy items. At St George's Hall in Liverpool a steam hoist was used by the local firm of Tomkinson Brothers (a firm which was in business until 1993). The machine ran on a tramway laid parallel to the wall and was said to accomplish its task 'with wonderful rapidity and precision'.[4]

A host of other steam driven devices for lifting, shifting and placing heavy components were devised in the middle of the century, gantry cranes, travelling cranes and a primitive form of tower crane. Machines were developed for excavating, for pile driving, for banking. New systems of scaffolding were developed and tried in a constant search for increasing erection speeds and greater safety.

It is true however that towards the end of the nineteenth and during the first part of the twentieth century, the British industry was lagging behind those of the USA and Europe. In America, the technology of building with steel and reinforced concrete was racing ahead. The problems of developing a successful pre-stressed concrete were solved in France but the material was for a long while used only intermittently in Britain.

There are two interesting accounts of the machinery in use in UK construction later in the twentieth century which seem to show that technical advance was taking place very gradually. In 1959, Dr G. Bonnell, the chief scientific officer at the Building Research Station (as it then was) listed the processes which had by then become mechanised or partly mechanised: joinery, plumbing (through pre-fabrication), the movement of materials, horizontally and vertically, and the digging of trenches and the mixing of large quantities of concrete. He claimed again that Britain had been very slow to adopt new techniques, citing as an example the use of tower cranes. He says the first were used in France in 1858 but that the first ones used in Britain were imported in 1950. (In fact, as mentioned above, there certainly were some equivalent machines in Britain much earlier.)[5]

Seventeen years later J.F. Eden identified all the main forms of mechanisation used commonly on sites in the UK. He did not show any major changes from Bonnell's list; and a similar list today would not be very different: tower cranes, other cranes, handling machines such as fork lift trucks, concrete mixers, pumps and a range of digging machines of every scale. Again the impression of slow, if any, change is difficult to avoid.

On the other hand equipment used on site today is very versatile and some of it is technically quite advanced. The 'JCB'-type machine can dig trenches, shovel earth, lift and carry heavy objects, knock things down and put things up. In fact so versatile is it that on relatively small sites it may be virtually the only large piece of equipment in use; it is its very versatility which makes this sort of machinery cost effective because the amount of idle time is obviously reduced to the degree to which it can be used on many different jobs. The same is true of the ubiquitous dumper truck, which can be seen carrying virtually anything around a site.

Britain may have been slow to adopt tall cranes but the ones used today are highly sophisticated pieces of machinery. They were designed 'with great skill and ingenuity' (to quote Eden) specifically for construction. The problems on site have often been not the shortage of the right technology but the planning of its use. As Dennis Harper pointed out in his excellent book on building production,[6] tower cranes in the early days were often more efficient as advertising hoardings than as lifting gear and as the accident described in Chapter 4 showed, there is still room for improvement.

So although the basic types of machinery have changed little since the 1970s, there have been constant improvements in their quality and capacity and the range of products on the market is so great that if a job on site can be assisted by mechanical means there is likely to be a machine ideally suited for the purpose. For example, in the 1990s there were approximately 14 manufacturers of truck-mounted cranes, each producing up to ten different models. There are some twenty makes of hydraulic excavators and a total of nearly 150 different models.

It is also worth mentioning here, as an example of the industry's capacity to adapt technology to its actual requirements, the development of the plant hire industry. Most of the equipment described above is extremely expensive but often only used for a short period of time or for intermittent periods on a particular project. The existence of a plant hire system allows the contractor to pay for machinery only when he needs it (though this is often quite a difficult planning problem) and allows the machinery itself to be hired out for different projects, kept in more continual use and therefore more likely to make a return on its capital cost.

As well as these large items of equipment, workers on building sites today have the use of purpose-designed power tools for virtually every job – drills, saws, sanders, staplers, and hand-guided hydraulic tools, mixers, and sprayers. The use of such tools is now so commonplace that it is easy to forget that this was not the case even twenty years ago. A study carried out in 1987 showed that sales of hand tools tripled in the previous few years, and though some of the increase was due to the growth of the DIY market, half of all sales were 'professional'.

There is also now a growing adoption of IT equipment for a range of functions – though again the industry has been accused of responding slowly to the possibilities. For example in 1996 a research group, Construct IT, surveyed 11 major contractors and came to the conclusion that they were well behind engineering firms in the use of IT. But things have moved ahead since then. In 2000, the contractor HBG was reported to be using a powerful electronic data system which could handle all the data and communications systems needed on very large projects – including the production and circulation of thousands of drawings; the firm was hoping to link up to its advanced 3D modelling system. (It might however be pointed out also that HBG is a Dutch owned company.)

Some of the early hesitation to use IT extensively was justified. For example, when architects were attacked for their reluctance to take up computer-aided design, the hardware and software were primitive by today's standards and difficult to use. Now that it is relatively easy to produce designs and estimates and use construction programmes with three-dimensional 'walk through' images, CAD has been taken up with great enthusiasm. The question of IT use is referred to again later in the chapter.

Despite the many technical advances, the Egan criticism may still be valid: that new technology is no good if it is simply used to 'reinforce outdated and wasteful processes'. The implication here is that the way buildings are built by bringing materials to a site and then putting them together is fundamentally inefficient even using modern machinery. One alternative is to move as much of the process as possible off the site altogether; the actual site of the building is simply where the final assembly of many pre-manufactured components takes place. This has all been said and tried before; some previous and current attempts to industrialise the construction process are described below.

Innovation, technology and the final product

However, first there is a very important further point to be made about technology and construction. It may be true that the level of technical sophistication on the average site for the average building might appear to be low but when we consider the level of technological innovation in the design and construction of the most advanced buildings today, we find a totally different picture; enough to claim that, at its best, the *level of technical sophistication is as high as in any other manufacturing industry*. There are so many examples that it would take many pages to justify such a statement fully. A few should be enough to make the point.

No. 88 Wood Street in the City of London, designed by the Richard Rogers partnership with Ove Arup as engineers is, like all their buildings, high tech from top to bottom; the glass cladding is made of a special super transparent low iron glass (manufactured in France); the lifts (made by Mitsubishi in Japan) were said to be 'the last word in precision engineering'; the huge triple-glazed panels are the biggest ever made. But these individual features are only a part, the whole building is as precisely engineered and detailed as a factory product.

At the British Museum, the new roof, designed by Norman Foster and Partners, consists of six thousand steel sections, each one different and manufactured to precise tolerances in Austria using computer-controlled cutting machinery. There are over three thousand differently shaped, specially-made glass panels, which filter UV light and control solar gain. Putting the whole lot together was a major technical triumph – the tolerances between structural nodes were 3 mm.

At the Eden Project in Cornwall Nicholas Grimshaw used a geodesic dome structure, not quite as complex as the British Museum roof, but hardly child's play; the glazing used a relatively new material – triple-layered ethylene tetrafluoroethylene (ETFE) – which is one hundredth the weight of glass.

There have been dozens of other 'high tech' buildings produced over the last decade or so, which represent a considerable achievement for British architects, engineers and contractors. They are mostly special landmark buildings and it is perfectly true they do not represent anything like the average. But most simpler and smaller buildings incorporate modern materials and well-designed, highly engineered components that have been constantly improved such as the modern domestic central heating boiler. Technical advances tend to filter down ultimately through the industry; perhaps a good example is the use of photovoltaic cell technology (which will be discussed further, in the next chapter); while its use was rare only a few years ago it is now being applied in projects of all types and sizes.

There can be no doubt that there has been and continues to be plenty of 'product innovation' in construction; whether it is widespread enough or fast enough is a question for discussion, but it is certainly true that buildings *are* being designed at the very edges of construction possibility.

It might be argued, however, that the kind of innovation seen in the most advanced building designs is not what the Egan recommendations are about. They imply standardisation of buildings, but each of the buildings mentioned above and the many others like them are unique and highly original. The constant search for novelty, the architect's desire to create something new and exciting might be seen as the precise opposite of what is needed. However, in each case there is a high degree of prefabrication and the construction process is a skilled assembly operation. Cost and time budgets are no more likely to be exceeded than with more standard buildings and client and public satisfaction is high.

Whether the industry can and should though be moving towards far greater levels of standardisation, towards the 'generic building type' with variations that Egan suggests is explored further in the next section.

Standardisation and prefabrication

Standardisation and prefabrication are not the same but are usually part of the same process. Most, but not all, prefabricated components are standard; most standardisation, but not all, requires prefabricated components. There are obviously different levels of standardisation and prefabrication. At one end there are, for example, items like bricks which are, after all, 'prefabricated' and, in shape at least, about as standard as you can get. At the other end there are elements such as prefabricated bathroom pods delivered complete with plumbing fittings which can be craned into position and quickly connected to the pre-installed services, electricity, water and drainage. We have only seen small examples of prefabricated complete buildings – such as the large 'mobile homes' permanently located on caravan sites (much more common and elaborate in the USA than in Britain). No one has yet parachuted a ten-storey building into place, though the American architect Ron Herron did design in 1964 a walking city that

Photos (10a) and (10b)   Barns: simple buildings but a revolution in technology – and in a way of life; (10a) 1860s; (10b) 2001.

was meant to move on gigantic legs from one place to another. That however was more science fiction than architecture.

Prefabrication can take place without standardisation. The elements for the British Museum roof quoted above were prefabricated off-site, but were all different, made individually (though they were in this case made using the same machine – an example of 'agile production'). Standardisation, on the other hand, does not necessarily require prefabricated components; there are a whole range of activities and procedures which can be standardised: delivery, storage, documentation, communications and, most importantly, design.

In the following sub-sections the two processes, standardisation and prefabrication, are considered together – because in practice that is how they generally occur – but the distinction should be kept in mind. The major issues and brief histories are considered under two headings - *housing* and *other forms of building*. Infrastructure engineering will not be discussed although it is another area where standardisation is highly important.

Standardisation and prefabrication in housing

Early examples

Some forms of standardisation and prefabrication have been normal in housebuilding for centuries; housing is the one area of construction where economies of scale can be achieved as large numbers of virtually identical products can be marketed. (The Egan report was not original in noting this.) In the eighteenth and nineteenth centuries, housing of all sorts was built from pattern book designs or simply to basic models developed by builders themselves. A historian of the period described 'the erosion of local vernacular styles by mass-produced, railway-transported components' during the second half of the nineteenth century, and the standardisation of the working class house was reinforced by the building by-laws which in effect produced the miles of terraced housing that still form the inner suburbs of many British cities.

In fact the massive housebuilding demand of the rapidly growing cites in the nineteenth century called not only for standard designs and repetitive processes but could not have been met without the availability of standard ready-made parts. As the demand was so high these – and this is not a contradiction – came in great variety (see Fig. 8.1).

From the beginning of the local authorities' building programmes (a small number from the 1890s but most subsequent to 1919), successive governments pressed for or insisted on standard designs and persuaded

CHAS. W.M. WATEFLOW
MANUFACTURER

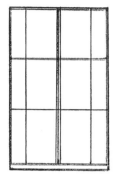

Of Sashes and Frames and Joiner to the Trade, 121, Bunhill Row, Finsbury-Square, – Well-seasoned materials, superior workmanship, lowest prices – Upwards of 400 DOORS and a large variety of sashes and frames always on sale. Glazed goods securely packed for the country. Steam-struck mouldings in any quantity – N.B. This establishment is worth the notice of all engaged in building.

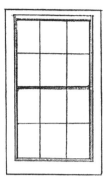

Fig. 8.1   Advertisement from 1845.

local authorities to experiment with new constructional methods. Even before the end of the First World War, the government had set up a Standardisation and New Methods of Construction Committee to examine alternative methods of construction and the standardisation of fittings. As materials and labour were in desperately short supply after the war, local authorities took up the ideas with enthusiasm. In Liverpool, for example, the council looked at a range of systems – steel frame, 'steelcrete', panel construction, and others, but in fact they used only two of these. None produced the expected savings in labour, time, money or improvement in quality and the city returned to conventional brick construction (to a limited number of designs) for their main building programme.

Two American experiments of the 1930s and 1940s are particularly worth discussing briefly because they illustrate very well the application and problems of standardisation and the mass production methods in house construction. The first was an attempt by the remarkable Buckminster Fuller (inventor of the geodesic dome) to produce completely prefabricated houses in factories. His Wychita house was designed to reach price levels below anything currently being built – but would require a production run of 500 000 a year to do so. Although prototypes were built the full project was never realised as no one was prepared to finance the building of the production factories. The designs were way ahead of their time in aiming for an absolute minimum of materials and energy use. There could hardly be a clearer example of the economies of scale problem in construction.

The second American experiment was more successful. During the 1940s a young entrepreneur named William Levitt built two new vast housing estates, one in Long Island and one in Pennsylvania (both of which he modestly called Levittown). They were built explicitly on mass production, 'Fordist, principles' and were meant to do for the house what Henry Ford had done for the car. Levitt explained that he had simply copied the production line system but whereas in a car factory the components moved to the men at Levittown the men moved along the production line. At the beginning there was one design, a single-storey ranch-type made from timber frame and plywood panel with felt tile roofs. Levitt divided the assembly process into 19 separate operations, allocating a small gang of men to specialise in each. House plots were spaced and laid out in long curving lines. Material was delivered with timber ready cut at the right time from Levitt's own mill in California; the gangs of men, each with their own specialised tasks to perform, moved smoothly from plot to plot. It is interesting that both the notion of the integrated production team and just in time delivery were applied here years ahead of their use in the Japanese car manufacturing industry.

For a long time, the houses looked like little boxes planted on a potato field – which is what they were; but the trees grew, people extended their homes upward and outward and planted gardens until today the estates have become attractive communities which are still much loved by the residents. Levitt continued to build similar houses on a similar system, this time in Florida, in the 1980s and 1990s and many of the people who had moved into the first houses in Long Island and Pennsylvania now retired to the Florida homes. Levitt was said by some to have, with Henry Ford, 'created the American dream' – bringing house prices within the reach of ordinary working people (as long, it should be said, as they were white; these were segregated communities).

Post-war Britain

In Britain, work during the Second World War on how houses were to be built in the future included studies of American experience and many strange ideas were considered, including a house designed by an American, Wallace Neff, which was built by blowing up several large balloons and then spraying concrete over the outside; the prototype looked interesting but nothing came of the idea. However in 1945, the first 30 000 temporary prefabricated houses were imported from the USA. The prefabs, which were meant to be temporary, provided good, if small, homes for many years and there are still a few left.

The housing drive of the early 1950s saw central government issue volumes of advice and directions to local government on efficient house-building and in 1952 the Ministry of Works, under Harold Macmillan, published a design for the 'People's House', to be used by local authorities throughout the country. These were quite a success but within a few years the drive for better, faster, cheaper construction led to a positive frenzy of experimentation in 'industrialised' or 'system' buildings, as the process was then known.

In a booklet published in 1964, over 32 prefabricated systems for low-rise housing were listed, 15 for high-rise and 15 for education and health buildings. Some were developed in the UK, others imported (Camus from France, Jespersen from Denmark, Sundh from Sweden and many more). It is difficult now to realise just how enthusiastic people were for the new systems, from central government, the construction industry, the architects, the planners and local authorities, but enthusiastic they certainly were. After the usual lament about building not having changed for hundreds of years unlike all other consumer industries, the preface to another 1964 publication includes this:

> 'But this is changing now. System building is bringing this industry into the machine age and will enable us all to live in clean and healthy sur-roundings, pleasing to our aesthetic feelings, at a cost which will be within everybody's pocket.'[7]

All the big cities followed the same path and built estate after estate of high-rise dwellings. To use Liverpool as an example again, the city con-tracted with the French owned company Camus to produce 2500 dwellings. A special factory to produce the units was built and the expected savings of 10% on conventional building were to be shared between contractor and client. Camus built in other cities but there were many other systems and the industry made a major contribution to innovative programmes. The results of the system-building boom have usually been described, in popular accounts, as disastrous. They were not all, in fact, as bad as they were made out to be but many turned out to be more expensive than con-ventional housing, riddled with faults and unpleasant to live in. Some of the high-rise system building, however, was very much better and in many cities the flats built at that time are now saleable at high prices. However the enthusiasm for high-rise industrialised housebuilding disappeared as quickly as it arose; the graph in Figure 8.2 tells the story.

There are many answers as to why this disillusionment happened in Britain and not in other countries. Living in flats has never been as common in Britain (with the exceptions of Glasgow and London) as in the continental

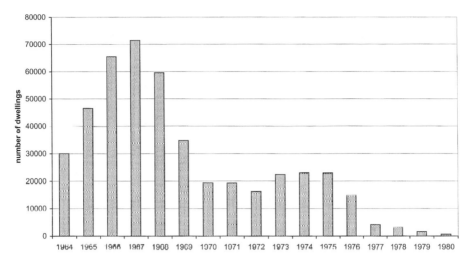

Fig. 0.2 Industrialised dwellings: approved tenders 1964–1980.

European countries and the poor quality of some of the new developments did nothing to improve the image of flat dwelling. Changes in the subsidy system made the true costs of building high more evident and they usually turned out to be much higher than for low-rise, but the critical event which is usually claimed to have sealed the fate of the industrialised housing programme was the partial collapse after a gas explosion in 1968 of a block of system-built flats known as Ronan Point, killing three people and injuring another 12. The cause was identified as a weakness in the load-bearing panel structure.

With the decline in all local authority housebuilding as well as high-rise, the use of other system methods also declined. The housing associations, which gradually took over the 'public sector' role, generally used conventional construction techniques and traditional procurement methods.

In some ways private housebuilders carried on as before, building semi-detached housing estates as they had done in the 1930s, but there were considerable changes taking place which included greater use of prefabricated parts – such as roof trusses, window frames, doors and staircases. More fittings and fixtures were included – complete kitchens and central heating. Competition for the new mass owner-occupier market led to continuous search for faster cheaper methods – often at the cost of quality, in spite of building regulations and the efforts of the National Housebuilders' Federation to establish minimum standards.

Photo (11)   Retreat from industrialised building – local authority housing from the early 1970s, (background) and the 1990s.

Many housebuilders have looked to prefabricated timber-framed systems to achieve higher construction speeds. Actual cost per unit can be as high as brick and block building but the reduction in time reduces interest charges and brings in revenue more quickly. For many reasons which would take too long to discuss here, timber-frame has had some difficulty in becoming accepted in Britain and some builders have actually reverted to traditional methods; however, one of the consequences of the new insulation regulations discussed in the next chapter, together with pressure for more effective prefabrication is that builders may more readily turn to timber frame in the future.

Two recent schemes

One of the most promising schemes currently getting under way is the work of the Amphion consortium. Originally three but now 20 housing associations linked up with housebuilder Beazer and the new housebuilding research centre, Zethus, to study and apply modern approaches to the large-scale production of timber-framed housing. By the end of 2000, three architects had produced designs, using the same basic Beazer system, for 17 different sites. The aim is to develop 2000 homes over a four-year programme.

One interesting aspect of the scheme is that the standard system can be developed by the architects into a wide variety of final buildings; it neither impedes architectural creativity nor suffers the problems of separation of the design stage from the construction stage. The approach to the complete process from manufacture to completion is based on teamwork between all the participants. Training is available for the actual assembly teams for this and other prefabricated systems from the Zethus centre. At the time of writing it is too early to see how successful this initiative will be; or to tell what will be the effect of Beazer's takeover by Persimmon, which seems less sympathetic to timber-frame.

A second well-publicised and apparently successful scheme in recent years has been the 30 flats built for the Peabody Trust at Murray Grove in London and claimed to be the UK's first multi-storey housing using prefabricated volumetric modules. The modules, each 8 m × 3.2 m and fully finished internally were built at Yorkon's factory in York and delivered to the site, two at a time, by lorry. They were then lifted into position, bolted together and linked to power and drainage, creating a block of complete flats five storeys high. A report on the project a year after it was built found that tenants were delighted with the homes, no defects were reported and no maintenance was required in the first six months.

However the capital costs turned out to be at least 5% above what might have been achieved through conventional building in spite of the fact that the modules were manufactured by the most experienced firm in the business (Yorkon is part of Portakabin) and erected by a Japanese company with specialised experience in this sort of construction. The extra cost was attributed partly to special bespoke elements required by Peabody and partly to the fact that finishing was not carried out in full production line conditions.

Standardisation and prefabrication in other sectors

Housebuilding is an area where the advantages of standard techniques and a high degree of prefabrication *should* be achievable. There are other building types that are required in more or less standard forms for similar purposes and here again there have been many experiments in systems approaches. Two of the most successful, which unlike many, produced

Photo (12)   Placing the prefabricated units at the Peabody Trust's new development at Murray Grove, London.

some very attractive results and buildings that served their users and purposes well were, the SCOLA (Second Consortium of Local Authorities) and CLASP (Consortium of Local Authorities Special Programme) systems. They were developed mainly for schools by consortia of local authorities. CLASP was also used for fire and ambulance stations as well as offices and houses on a modest scale.

Both systems were based on light steel frames (rigid in the case of SCOLA, pin jointed in CLASP) to which various prefabricated cladding elements could be fixed. Everything was designed on a modular basis so that components could be fitted together in a variety of different ways and both systems were said to be economical to a height of four storeys. Although the buildings built under these systems have recognisable similarity, they were capable of producing in the hands of imaginative architects individual and interesting buildings; at least one school won an international architectural prize at the Milan Biennale.

CLASP was used for office buildings and housing as well as schools but it was not designed for more that two storeys. High office building, which in this country did not begin to approach the skyscraper ambitions of the USA and other countries, needed a different approach, but here also there was a high degree of prefabrication and standardisation.

It is easy to underestimate the degree of prefabrication which has become standard practice in the construction of office buildings. Concrete and steel frames with prefabricated cladding systems were normal for most office building from the late 1950s onwards. Many of the industrialised building companies that tendered for the local authority high rise housing projects of the 1960s had in fact developed their systems in rebuilding commercial centres over the previous ten years. The complaint was not of lack of standardisations but of faceless, repetitive, concrete commercial architecture. In the late 1980s there was much discussion of 'fast-track' building methods as exemplified in the massive Broadgate development at Liverpool Street station in London.

Fast track was essentially a management contracting procurement method allowing design and construction phases to overlap but it involved also using high levels of prefabrication and rapid mechanised assembly. At Broadgate, prefabricated cladding, interior wall panels with radiators, toilet pod modules, modular lifts and plant rooms all contributed to the speed of erection. The actual techniques of installation were also standardised and accelerated.

This was a speculative development for multiple clients but it is the large single clients with regular building programmes that have developed the most standardised approaches particularly the major retail and restaurant chains. McDonalds, for example, use contractor Britspace's modular system for constructing restaurants and have recently developed the modularisation further for the roofing structures.

An interesting new venture outside the commercial sphere and with some similarities to CLASP is the 'Optimum Sports Hall'. It is to be offered by Sport England as a standard package of design drawings, ready to be used as tender documents. There are three basic designs, using a braced steel frame with a standing seam aluminium roof – but the designs can be varied without losing the advantage of rapid and cheap construction. The system was developed by a team that included architects, structural and service engineers, cost consultants and buildability consultants. Similar standardised swimming pool designs are expected to follow soon.

There are several intriguing aspects of this project: it seems to overcome the separation of functions between architect, engineer and builder, without sacrificing design quality or buildability; the cost of the buildings should be highly predictable and the procurement route could take several different forms.

However, despite all the advances made towards greater standardisation, it is clear that the industry as a whole is nowhere near approaching what Egan calls the 'generic building type' – a kind of universal standard (like the car body) which is then varied to suit many different functions and locations. Whether this is desirable or achievable remains an open question.

Coordinating the process: managing the supply chain

The third route to greater efficiency identified at the beginning of the chapter was effective control of all the inputs to the construction process. It is in this area that the building and construction business has notoriously been weak. Whenever buildings have overrun their budgets or construction periods the reasons are nearly always failures at the interface between one stage of production and another. Delays in the completion of one stage create backlogs, wasted time and unmet contractual conditions. At any stage delays can be caused by failure of materials to arrive, inadequate or impractical design detail – or last minute alterations to design – and a host of other problems nearly all of which are to do with the links between groups of people with separate responsibilities. Sometimes the whole construction programme breaks down because of disputes – over contract conditions, methods of working, payment, standards of workmanship, division of responsibility and many differences of perspective.

These kinds of problem are so common that Egan was obviously right to concentrate on ways of overcoming them. They have always been part of construction, but this does not necessarily mean they have to be with us for ever. Two of the major sources of the problem have been discussed in detail already in different contexts – but we look here at their consequences (or claimed consequences) on site. They are:

● the buildability problem resulting from separation of design and construction
● the extent and nature of subcontracting.

Two others have not been discussed before:

● inadequate planning of operations
● difficulties in the supply of materials.

These last two are discussed first below, before we take another look, in the context of the site operation, at buildability and subcontracting.

Planning the process

Inadequate planning should no longer be an issue but there are apparently many contractors, particularly the smaller ones, whose planning is still done in a rough and ready way and who rely on the casual phone call for fixing dates with subcontractors and materials suppliers. Planning *should* not be a problem because for many years now the whole subject has been carefully investigated and widely taught on all construction courses. The then Ministry of Works published a booklet called 'Programming and Progressing for Housebuilders' as far back as 1941; it was hardly the ideal time but it was followed by a stream of advice leaflets over the next 20 years. There have been countless publications on the main methods – from the simple bar charts used in the early days which some builders still find adequate to the more complex methods using critical path methods such as PERT (Program Evaluation and Review Technique) and what is today called process mapping.

It is now possible to integrate the whole design and construction pro gramme using computer-aided design data linked through to the programming of actual construction, with information on progress and delays instantly leading to revised optimised alternative sequencing. In spite of the examples such as HBG quoted above it still seems to be quite rare for contractors to exploit the full potential of IT here. That potential, however, had been well understood by some for quite a few years as this quotation from the director of housebuilder Westbury in the mid 1990s shows:

> 'What we are really doing is turning this business into a manufacturing operation.' The claims made were impressive: 'QS measurement time per house has been cut from 25 man days to 15 minutes, the 51-day process to design plan cost and issue orders for a house cut to about 21 days; the exact amount of materials required is sent to site at exactly the right time, mimicking the just in time system delivery systems of the Japanese car manufacturers.'[8]

(This is, one might add, exactly as Levitt was doing 50 years before.)

What is required now is the development of only a very limited number of systems which can be easily taught and used even by contractors operating on a moderate scale; the problem seems to be that at present no-one can decide which of the systems on offer is to be the standard in the long run.

There have also been long running problems with the management of production at site level. Dennis Harper wrote in 1978 that 'the organisation structure required for site control and supervision of a large building project remains confused and rather obscure'.[9] Egan identified training at the supervisory level as one of the key weaknesses of the current system. It is clearly no good having the most sophisticated computer-based planning systems if there is not high quality supervision at site level to ensure that things actually happen more or less in the way they are supposed to.

Whatever the sophistication of the planning it will never be foolproof and will never work well without coordination of planning throughout the supply chain of subcontractors and suppliers.

Materials and components

Most building projects use literally hundreds of building materials though a few – concrete, steel and bricks, for example – will dominate in terms of bulk. Some of these materials will be supplied direct by the manufacturers, others will themselves have come through a long supply chain – from production to fabricators to factories and maybe builders' merchants. Contractors have always been aware of the desirability of having the materials delivered to the site as closely as possible to the time they are needed. Storing large quantities of materials on building sites is often difficult because of space, but also invites theft and damage (from weather, vandals and wayward dumper drivers); it is one of the main causes of waste usually estimated to be around 10% on the average building site. Again, most contractors do not need official reports to tell them what the problem is; what is needed is a feasible solution. Many reports including Egan have proposed answers. But first consider some of the difficulties.

Figure 8.3 illustrates one aspect of the problem that faces the contractor or project manager. It shows the so-called 'lead times' for supply of various building components at one particular time. It can be seen there that lead times can be very different for different elements of a building. That in itself raises a planning problem, but the problem can be made much worse by sometimes sudden changes in the lead times of different items.

The reason is the same phenomenon we have come across many times – the erratic nature of demand for construction – and the problem is made more unpredictable because many specialist materials are imported; it may be difficult to obtain the Spanish slate specified for the roof because of a building boom in Spain – or problems in the Spanish quarries – way beyond the control of the British contractor. (The possibility that this particular one could be solved by using Welsh slate or cheaper manufactured slate raises a host of other issues which cannot be explored here.)

The sort of headache this can create for managers is illustrated by the

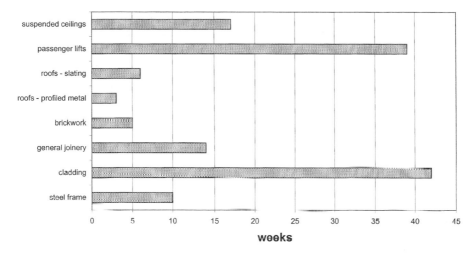

Fig. 8.3   Lead times (5-storey office block).

rapid rise across Europe in prices and lead times of glass in October 2000; prices of some grades were said to have risen by 50% and lead times to four months. One contractor said that glass manufacturers and suppliers 'were reluctant to increase capacity for what they see as short-term market demand'. The impact of rapid fluctuations in demand could not be seen more clearly.

Separation of design from construction

We saw in Chapter 5 that it is a common complaint of contractors, particularly under the traditional systems, that the design drawings and specifications they are asked to work with are often not clear, are sometimes subject to late variations and most importantly have elements that are not buildable. This so-called problem of poor 'buildability' is seen as the consequence of the fact that architects prepare designs in isolation and do not themselves have sufficient knowledge of construction detail. However, the problem can be exaggerated and should not really be too difficult to solve.

First, architects trained in good schools should in fact have a sound knowledge of the buildability of their designs; buildability is one of the major criteria used in critiques of student work (though some schools seem to have moved away again from this practical approach) and most quickly learn through experience – especially with angry contractors!

Secondly even under the traditional system, relationships between client, architect, engineer and contractors can be and in most cases usually are very positive; problems of buildability are frequently resolved at the regular site meetings between architect and contractor. Sometimes too a contractor's protest that something is not buildable turns out to be the result of a mis-understanding of the architect's intention; sometimes with discussion the contractor will suggest a different way to achieve the same result. There is no denying, however, that difficulties do arise and can lead to major dis-putes. Forms of procurement and forms of business organisation such as multidisciplinary practices that allow interchange of ideas at the early stages are obviously ways of overcoming the problem. However no form of procurement is going to eliminate the need for real understanding between the people with different skills, and real inadequacies on all sides have to be recognised and tackled.

A possible example of the way things can be done successfully is the aquatics centre in Manchester. It was procured using a *three*-stage tendering process which, press reports indicate, went a long way towards solving some of the problems discussed above. Interestingly this solution seems to have arisen from a conflict between the council's desire for a design and build approach to reduce risk and Sport England's worry that conventional design and build would not produce the right quality. A 'buildability consultant' (Laing) was appointed at the first stage to work with the architects, Faulkner Brown. A two-stage tender for a lump sum building contract, with the first stage selecting the contractor on a number of initial criteria and the second stage to 'procure the subcontractors and value-engineer the works packages'. The original architects were novated to the successful main contractor (Laing again). The project was completed ten weeks early – apparently to the satisfaction of most of the participants and clients.

Subcontracting

Many commentators on the industry have seen the really fundamental problem of the construction process – more significant than any we have discussed so far – to be the extent of subcontracting and the system of individual contracts on which it is based. Subcontracting is not unique to construction of course. Most firms in most manufacturing industries sub-contract much of their component production to independent specialist suppliers; what is different about construction is that *it subcontracts much of the final assembly process* itself. The number of subcontractors operating at one time or another, on a single site, can be very large on a major site. There were over 400 working on One Canary Wharf, and, even more astonishing, it was claimed that there were 700 different main trades involved with 1000 sub-trades.[10]

We saw in Chapter 6 how subcontracting in its modern sense developed as a fundamental element of the industry from the early nineteenth century; even though it was seen as rather disreputable. A 'respectable' builder was defined as one who did not need to subcontract. It was argued that the fact that subcontracting became universal against the opposition of workers and clients alike seems to indicate that there was some deep and perhaps unavoidable logic to the practice.

That logic is the same today as it was in the nineteenth century and the advantages of some types of subcontracting have been reinforced by the employment and tax legislation discussed in Chapter 4. Flexibility in response to variations in demand and in size of project is still important and the subcontracting system can still help to provide it. The development and growth of the subcontracting system was and still is a rational response to the many sorts of irregularity in the demand in the industry. For, as well as the overall problem of market instability – the slumps and booms which have been with us for centuries – there have always been the difference in types of input needed for individual projects and the varying, often inter-mittent, time periods needed for each project. However, the most important stimulus to the growth of subcontracting (other than the labour only variety) has more recently been the highly specialised nature of much of the work (which is also, incidentally, another indication of the industry's increasing technical sophistication). Again the figures from Canary Wharf illustrate this point; nearly all the subcontracting firms were specialists of one sort or another.

In fact subcontracting is an excellent example of the division of labour which has been seen as one of the underlying generators of economic efficiency. Although there are few scale economies in the production of buildings on site, there are scale economies not only in the production but also the installation of individual components across the industry as a whole. To take an example, it would obviously be utterly uneconomic for a medium-sized contractor that builds multi-storey offices to employ its own lift engineers, who would be used once for a short period on each contract. The specialist lift firms, on the other hand, can provide continuous employment to the engineers who will install lifts in projects over a wide geographical area.

However, there is no doubt at all that the subcontracting system has also been a major part of that fragmentation of the industry which has been discussed in so many different contexts before, as a principal cause of inefficiency. It is certainly the source of the expensive burden represented by the enormous cost of disputes. Failure of subcontractors to perform to required standards (which may not be their fault) or even business failure of the subcontracting firm itself can cause major disruption to building programmes. The construction of the building in the city of London referred to above was seriously delayed when the American cladding subcontractor went into liquidation half way through its £18 million contract.

Whatever the detailed reasons in this case, it does show that even at the highest level the existence of a formal contract is no ultimate protection against disruption. The alternative arrangements proposed by the Egan report, described in the last section below, would attempt to prevent these kinds of situation by establishing long-term relationships between contractors and subcontractors and greater openness of accounting. There probably is no absolute protection – which is why construction will always remain a high-risk business – but there may be ways forward towards greater control of the risks.

Subcontracting is a permanent part of the system, but the subcontracting sector of the industry may itself be far too fragmented. Apart from the issue of labour-only subcontracting which was discussed in Chapter 4, there may be a problem of too many small specialist firms that do not have the capacity to train and permanently employ skilled staff. In order for them to grow they themselves need some stability of work; always difficult to achieve in construction but again, something the Egan proposals may help to address.

Integrating the whole process: some solutions

Many of the reforms currently proposed have been debated and even attempted often before. The quotation from Cleeve Barr in Chapter 1 – with its reference, *nearly 40 years ago*, to the need for 'new working relationships between clients, architects, builders and manufacturers' – shows the Egan ideas are not themselves very new. Academics such as Professor John Bennett at Reading University had pointed out that British construction firms were learning from the Japanese 'the crucial benefits of co-operative long-term relationships between customers, designers, managers and specialist contractors'.[11]

There has been no shortage of analyses of the problem and broad specification of solutions; making it all happen seems to be the difficulty. The concept of 'partnering the supply chain' is really quite radical and requires new ways of thinking. It extends the idea of partnering between clients and contractors discussed in the last chapter through the complete chain of materials and component suppliers, subcontractors and main contractor. There is a new growing literature on this and no doubt much more to be said but in essence the proposal is this: contractors will establish, over a number of projects, long-term relationships with subcontractors and materials suppliers. This will involve 'open relationships, effective measurement of performance and an ongoing commitment to continuous improvement' (Egan para. 45).

The aim is a complete integration of the team – from designers through to the smallest subcontractor – using 'lean production' techniques and value engineering to identify and eliminate waste and to identify improvements throughout the whole process. Most importantly this integration should carry through from one project to the next, each job building on the experience of earlier ones. This has already been achieved by some big contractors on major schemes. For example, HBG worked on seven consecutive projects for the development company Argent Estates with what they described as 'a secure integrated team' enabling them to make improvements from one project to the next; they claim substantial reduction in costs and a 20% reduction in defects. Further standardisation and prefabrication are expected to lead to further efficiencies.

It is still too early to judge just how successful these proposals are likely to be. The Movement for Innovation (M4I) has been established which, together with the government's Best Practice programme, is aiming to preach the Egan gospel across the whole industry. The government is also using its own enormous influence as client to push the industry in the same direction. Prime contracting, discussed in the last chapter, implies the application of all these principles. Government departments are expected to change their procurement systems; housing associations are expected to demonstrate 'Egan compliance' in their development proposals. Information on all these and other developments is available on the internet and some website addresses are given in the 'For Further Study' section at the end of the chapter.

Underlying all the issues discussed in this chapter – achieving technical innovations, improving prefabrication techniques where they are appropriate, and effective co-ordination of the construction process – is the need for improved and perhaps a different sort of training and education. The research team at the University of Westminster, referred to in Chapter 4, came to the conclusion that one of the key factors in the capacity to innovate was the level and type of education of the workforce throughout the industry. This is recognised too in the Egan report, but it is difficult to discern whether any of the major changes needed have yet even been fully defined.

Yet if the advances required by Egan are to be spread *throughout the industry* rather than remain the preserve of the big clients and contractors, the obstacles to radical changes in the training system, including the lack of steady full-time employment, need to be overcome.

For further study

The concept of 'economies of scale' is more complex than it appears here. Most of the standard elementary economics texts discuss it in more detail and it would be worth looking at some examples and thinking this issue through further.

Part Two of Akira Satoh's *Building in Britain* gives a well-illustrated account of technical developments in construction during the nineteenth century. Dennis Harper's *Building – The Process and the Product*, although over twenty years old, is full of useful information and ideas. The section on site organisation is particularly relevant.

More recent books of direct relevance to the theme of this chapter include: *Construction Process Improvement* edited by Atkin, Borgbrant and Josephson and *Construction Project Information* by Aouad, Tah, Alshawi, Underwood and Faraj.

The journals are most important sources, particularly case studies of recent building and construction projects, which indicate how issues discussed in this chapter are overcome, or not overcome, in practice.

Useful websites include those quoted at the end of Chapter 7, but perhaps the best way in which to obtain information on the management issues of construction is through the website set up by the University of Reading Department of Construction Management and Engineering www.construct.rdg.ac.uk. It has links to a large number of other relevant sites.

9 Construction and the Environment

- Introduction: three dimensions of 'sustainability'
- Energy and buildings
  - Global warming and the economics of building
  - Insulation standard
  - Energy efficient systems
  - Passive solar design
  - Local wind generators
- Building materials and the environment
  - example 1: wood
  - example 2: uPVC
- The built environment: for better or worse?
- Where we are now

Introduction

No discussion of the industry today can be complete without some reference to environmental issues, to the notion that construction should become more 'green', 'sustainable' or 'eco-friendly' – the three currently used terms to describe what is basically the same objective. There has been much effort over many years by government and various pressure groups to make the construction industry more aware of the impact of building on the global environment, particularly through its use of energy. The industry's response has in the past often been sceptical but more recently there has been a major change in attitude; there is not only a more enthusiastic and positive response to calls for greater sustainability, but also real advances in environmentally friendly construction and design techniques.

Construction affects the environment in a host of different ways. The most frequently discussed is the large amount of energy used in buildings once they are built and occupied, but the construction process itself is highly energy dependent – particularly through the manufacture and transport of materials. There are more obvious impacts; construction work of all sorts requires space and alters the character of the environment in which it exists.

There are also indirect effects of the location of buildings – such as its influence on transport requirements, for example in increasing the use of cars (also high energy polluters) to out-of-town shopping centres and office developments.

There is now a large literature on the environmental problems themselves and on the specific effects of construction. There is a mass of academic and government advice as well as regulation; there have already been hundreds of buildings designed and built demonstrating how environmentally sound construction can be achieved.

So in one short chapter it is possible only to delineate some of the major issues, to describe some of the problems and their potential solutions and to look briefly at what has already been achieved and the directions of future advance. The chapter is divided into three main sections, looking first at the major question of the energy efficiency of buildings, then at selected examples of construction materials and the environmental implications of their use and manufacture and finally at the immediate physical impact of construction as a whole.

Energy and buildings

Global warming and economics

As scientists became convinced that global warming was a reality and that the cause was depletion of the earth's ozone layer through emissions of a range of carbon gases, several intergovernmental conferences were held to determine what action could and should be taken. After the Earth Summit at Rio de Janeiro in 1992 the Construction Industry Council in Britain set a number of goals for the industry including a 50% reduction of waste on sites. Five years later, the British government signed up to the Kyoto Protocol and the Prime Minister committed the UK to the ambitious target of a reduction in the emission of carbon dioxide to $22\frac{1}{2}\%$ below its 1990 level by 2010. Although many believe these efforts to be essential and in fact not ambitious enough there are dissenting voices – including the American government since the election of George W. Bush as president. These argue that the costs of drastic policies will be greater than their benefits and that the problem is not as serious as environmental campaigners have made out. However, their more optimistic projections often depend on the assumption that energy-reducing measures will in fact take place; the scientific consensus is that the global warming problem is real and is serious.

Whatever the final resolution of these debates may be, there seems to be no doubt that the saving of energy is an important objective; waste of energy is waste of resources which could be used for other purposes. In the case of construction the argument has been that it is in fact a massive and wasteful user of energy. The usual figures quoted are that construction (both the process and the buildings) is responsible for about 50% of the emission of greenhouse gases in the UK; but the more astonishing figure is that 60% of that is attributable to space heating of buildings. However, buildings use energy not only in space heating but also in the heating of water, in lighting, mechanical ventilation and machinery such as lifts. All this is why of course the industry has been the target of a concentrated campaign to reduce its energy consumption.

Using energy costs money; a reduction in the amount of energy used in buildings, houses, offices or whatever would bring considerable financial benefit to their owners and users. In 1997 the Construction Industry Council estimated that clients could save up to £4 billion a year by using energy efficient buildings. Why therefore has it seemed necessary for governments to impose regulations and to shower the industry with exhortations and advice to persuade it to change its ways? The answer is, partly, simple economics.

Most (but, importantly, not all) measures to reduce energy consumption also cost money; if this were not so, consumers would have been more vociferous in demanding more energy-efficient buildings and builders would have been able to use energy efficiency as an effective marketing feature. If the costs of adding on or designing in energy-efficient features in a building are greater than the value of the consequential energy savings, there will be no economic advantage for building owners or users. Unfortunately this has often been the case and even where there are economic advantages to consumers they are not always obvious – or may take a long time to materialise.

The basic economic problem is that establishing the relative costs and values is not easy. It requires the use of whole-life costing techniques, that is, it requires the comparison of an initial expenditure – an expenditure at one point in time – with savings accruing over a long period. The principles of making these calculations are well established. There are simple rule of thumb methods such as calculating how long it will take to pay off the costs of an initial investment in an energy-saving measure; and there are more complex techniques, which involve the discounting of future values at some agreed rate of interest.[1]

These techniques are dealt with now as standard in many surveying and construction texts and they will not be explored further here. The point is that as often as not when applied to energy efficiency measures they often show that there is very little or even no advantage – certainly in the short term – for users. Housebuilders, for example, have often argued that meeting high energy-saving targets would add more to the costs of the house than users are prepared to pay. Individual house owners who install solar-heated water panels on roofs in most parts of the UK will find that, however they do the calculations, the saving in fuel will not pay for the capital cost of the installation. (In other countries such as Greece on the other hand the pay-off is fairly immediate.) Again, the installation of double-glazing in an old house may actually cost more than the value of the savings in energy.

However, there is another side to this argument. As higher efficiency standards have come to be expected, their costs have simply been absorbed and treated as normal. Central heating, a luxury in this country only 30 years ago, is now standard in new housing. The cost is no more significant as an 'extra' than the cost of bricks – it is just part of the whole. The same is becoming true of double-glazing.

These two examples of change are generally due to rising incomes and expectations but other improvements such as better insulation in houses seem to have required at least government pressure and often regulation. In commercial buildings particularly, where the people who incur the costs may not be the same people who make the initial design decisions, external pressure may be even more necessary; though here again increasing awareness of the possibility of energy savings is making clients more demanding.

If we are to meet our greenhouse gas reduction targets, a continued programme of persuasion and regulation is probably still necessary. Persuasion is necessary partly because economic advantages to users are either not easily available or are not easily perceived but also because designers and builders have not always been aware of the possibilities. Some energy-saving can be made with little or no extra cost at the design stage. For example it may cost no more to put a given number of windows in the optimum position for energy conservation than in non-optimal positions. In the long run consumers themselves will gain the advantages of lower fuel costs and greater comfort – and will then wonder, as with central heating, why it was not always like this.

There are three principal ways in which reductions in the manufactured energy used in buildings can be and have been achieved, these are

- improved insulation to reduce heat losses through the building's fabric
- the use of energy-efficient heating, lighting and ventilation systems
- the use of 'passive solar design'.

The next sections touch on each of these in turn.

Higher insulation standards

One obvious way of reducing the amount of energy needed to heat a building is to keep the warmth inside for as long as possible. If outside temperatures are lower than those inside heat is bound to dissipate ultimately and if windows or doors are left open it will disappear very quickly. It will also escape through the fabric of the building but the rate at which it does so depends on the thermal conductivity of the material from which walls, windows, roofs and floors are made. The reduction of heat loss by using or adding materials of low conductivity – that is by better insulation – has been the principal way in which energy savings have been sought over the last 35 years.

Part L of the Building Regulations, the section dealing with conservation of energy and power, was introduced in July 1965. It included for the first time specific requirements for the insulation of buildings. These were progressively raised during the 1970s, initially in response to the rapidly increasing price of oil. The major objective has changed to the prevention of emissions of CO_2 but the basic intended effect of the regulations remains the same – the reduction of energy used in buildings.

A full understanding of the way the regulations work requires a study of Part L itself but the basic principles can be outlined without too much distortion. The insulation requirements are set out in a number of different ways for residential and non-residential buildings. Common to both are target 'U-values' for different elements of the building.

U-values measure (in watts per square metre per degree Celsius) the rate of heat loss through the fabric of the buildings. The lower the U-value, the lower the heat loss and therefore the higher the insulation effect; a wall material with the unlikely U-value of zero would allow no heat to escape at all. In fact the current (post-1994) requirements for walls is a maximum U-value of 0.45 (that is, a wall must lose no more than 0.45 W per square metre per degree C). In the case of residential buildings, the actual U-values of some elements depend on other energy characteristics of the building measured by its Standard Assessment Procedure (SAP) Energy Rating. A building with an efficient heating system will have a higher SAP rating than one with an inefficient system, for example (other things being equal). Table 9.1 shows a selection of required U-values applicable before and after 2002.

New regulations proposed in 2000 require a lowering of U-values for house elements up to 2003 (see Table 9.1), and although in numerical terms the actual figures seem to be insignificant their effects on the industry could be quite dramatic. In fact, so concerned were many housebuilders at the original proposed changes that they persuaded the government to relax

Table 9.1 Examples of changes in thermal efficiency requirements for houses

Element	U-values for houses with efficient heating	
	pre-2002	post-2002
Pitched roofs	0.25	0.2 (0.16)*
Walls	0.45	0.35
Ground floor	0.45	0.3
Windows	3.3	2.0

*depending on insulation; 0.16 if between rafters

them a little. The proposal for walls was to reduce values from 0.45 to 0.3
(see Table 9.1). But this has now been changed to an initial reduction to 0.35.
The reason for the concern was that it was believed that the level of 0.3
would be impossible to meet using conventional block and brick
construction. Housebuilders thought they were going to be forced to make a
massive shift into timber frame construction. There has been considerable
argument over the whole issue. Block manufacturers have claimed the
lower U-values could in fact be met while on the other hand the housebuilders already specialising in timber frame construction have accused the
government of weakness for backing down in the face of pressure.

Another major implication of the new regulations will concern window
manufacture and installation – particularly as they might apply to replacement windows in old property as well as new-build. The proposed
reduction (by 2004 but this may change) is to $2.0\,\mathrm{W/m^2/^{\circ}C}$ as opposed to
the current 3.3. This would require all windows to be double-glazed and
metal windows to be triple glazed, possibly requiring the use of coated glass
on the external panel. There are limits to a designer's ability to overcome the
heat loss problem by reducing the percentage of glazing because the
regulations also require sufficient glazing to allow adequate daylight.

There has already been an enormous improvement in the energy efficiency of new houses. Double glazing, loft insulation and central heating
have all become standard in new houses and been installed in millions of
older ones, but to achieve the new standards may require more radical
rethinking of design and construction methods. It could well increase the
costs of houses and other buildings considerably, and there are bound to be
questions raised about how far these costs are to be borne by the individual
building users and how far they are perceived as a contribution to national
objectives and perhaps supported through subsidy of some sort.

Energy-efficient systems

High levels of insulation prevent the warmth generated in a building from escaping too easily. Energy can also be saved by ensuring that the heat is produced in the most efficient way and that all other energy demands in the building, such as for lighting, water heating and ventilation, are also met with minimum energy costs.

Lighting and heating systems can be energy-efficient on three different levels: they can use the most efficiently produced energy, they can use the form of energy most appropriate for the purpose and they can convert the energy in the most efficient ways.

Lighting

In the case of lighting (and ventilation), there is today no real choice in terms of energy type – it will almost always be electricity. The electricity may itself come from more or less efficient and more or less environmentally destructive forms of generation. It would take too much space here to investigate the complicated issues of comparison between coal, oil, gas and nuclear electricity generation, though we look briefly at wind and solar power below.

Wherever the electricity is generated, it can be converted into light more or less efficiently. At the domestic level, low-voltage high-efficiency light bulbs are becoming more widely available and cheaper, but the decision to save energy by using these where appropriate is a decision for the individual householder, not the builder. This is a good example incidentally of an area where the problem is one of awareness rather than of cost. As prices of low-energy bulbs come down, they have become more cost-effective with a reasonably quick payback in electricity savings. Yet they are still perceived as expensive and take-up is still very low. The same problem of awareness is evident in the failure to use the most effective of all ways of saving electricity in lighting – which is to turn lights off when not in use. Here other conflicts arise; for example, people prefer working in buildings which are well-lit all over, all the time. The bright lights of the city are a central feature of urban life.

In large commercial buildings significant energy savings can be induced by the design and choice of lighting systems. There are, again, potential conflicts between the short-term economics of a system from the user's point of view and global energy saving potential. However, a study published in 2000 shows that if users are prepared to consider whole-life costs when making decisions there are some clear benefits to be gained by making the right decisions.[2] The study compared seven different lighting systems for a large open-plan office. Two systems produced lower energy costs than the others; in fact only half the cost of the most energy-expensive, but, in purely financial terms, one of these efficient systems works out as the most expensive both in terms of the initial investment and in terms of whole-life costs. The other energy-efficient system has, however, one of the lowest whole-life costs of the seven; furthermore the cost of the lamp technology used is reducing relatively. So although, as the study points out, the design of systems for large offices is a complex business and there are many other considerations, it does demonstrate that considerable energy savings can be made by careful examination of the alternatives available.

Heating systems

In the case of heating there is both a choice of fuels and a range of energy conversion efficiencies. The question of which fuel is 'best' has always been complicated for designers and users by changing relative prices of the different fuels; in a perfect market system the relative prices should reflect the relative efficiencies of production but energy markets are not at all perfect. Sometimes oil has seemed the cheapest central heating fuel and at other times gas. Oil prices, for example, are often manipulated by the oil-producing countries to maximise revenue over time and prices may not reflect the real economic costs of extraction and distribution.

What does seem to be generally accepted today is that the use of electricity for heating is almost always a poor choice both from the point of view of 'global' energy use and from the point of view of the individual consumer. The reason is that at every stage of conversion, from say oil-fired power station to the emission of heat from an electric radiator, most of the energy is wasted in some form; as a result the proportion of the primary energy released as heat into a room is very small. Storage heaters that use energy produced at times when demand is low help to increase the overall efficiency of the generation process but that effect may well be counteracted by the weakness of control over the release of heat from the storage units. Again, as in the case of lighting, electricity from environmentally friendly sources such as wind power will reduce the disadvantages of using it for heating but the conversion process to heat will still be inefficient; such electricity is more effectively used in driving machinery, in lighting and in electronic equipment.

The other main possible fuels for heating are coal, oil and gas. Coal is still used in some large central heating systems and, of course, still in some homes though the use of coal as the main domestic fuel is now minimal. The real savings to be made are from the use of efficient oil and gas boilers. Here, the improved technology of boiler manufacture is already yielding significant results – but builders and householders still need to be made aware of the possibilities.

For example a domestic gas-fired condensing boiler produces high-energy efficiencies by recovering heat from the flue gases. Even when their installation costs were twice the costs of conventional boilers they yielded positive returns in a relatively short time. As they become more common and relative costs reduce they will be even more worthwhile – for the consumer as well as in terms of their contribution to lower energy use generally.

Passive solar design

All of our energy is ultimately derived from the sun, it is the sun's energy which is embodied in oil, coal, natural gas and timber, all of which has to be expensively converted into useful forms. But the sun's energy is also available freely as natural daylight and direct heat radiation. Obviously, the more use that buildings can make of this freely available source of energy, the less need there is for manufactured forms. The various techniques for tapping this natural source are often referred together as passive solar design and include:

- Optimising the use of windows and glazed areas
- Using the building's thermal mass as a form of heat storage
- Using solar thermal systems such as heating panels
- Using natural ventilation
- Using photovoltaic cells, which convert the sun's energy directly into electricity
- Using local wind generators
- Using the thermal mass of building elements as temperature stabilisers.

A few examples only are discussed below.

Optimising the use of glazed areas

This often presents a complex design problem but the principles are now well understood and there is no longer any excuse for ignoring the issue as was often done in the past. The complexity arises from the fact that glass allows both light and heat to pass through. Large areas of glazing increase natural light levels but can create glare; they can produce too much heat from solar gain during warm and sunny periods but lose too much heat when it is dull and cool. Double- and triple-glazing and the use of special glasses can reduce heat loss – but also reduce light levels. As new forms of special glass and other substitutes become available this problem may be less significant. Several new buildings, including the Eden Project in Cornwall, referred to in Chapter 8, use a material known as ETFE (ethylene tetrafluoroethylene), which is translucent, recyclable and one hundredth the weight of glass.

Using conventional glazing, however, it can be difficult to get the balance right. Designers today use a combination of different approaches. The shape and orientation of a building and the distribution of spaces are obviously significant considerations. A position which allows the maximum heat gain in winter and the lowest in summer may not be achievable in dense urban environments where other buildings and existing building lines limit the possibilities, but on new sites a great deal can be achieved – just by thinking about the problem. It is obvious that on many speculatively built estates there has been very little consideration of this sort; best solutions can be difficult to find and need careful design. On others such as the Millennium Village project, for example, all aspects of the fenestration, the manufacture and design of the frames, the type of glass, the size and orientation of the windows, were calculated to make the best use of the sun's heat and light energy.

There are many other design features which can inhibit or maximise the use of natural light while reducing heat gain. For example, the avoidance of deep plans or introducing daylight through light wells, roof lights and glazed atria (all of which have been used for at least 200 years); light can be reflected internally and the maximum use made of borrowed light from one working or living space to another. Many modern buildings are now designed with shelves – sometimes adjustable – above the windows to provide protection from excessive solar gain when the sun is high in summer but maximise solar gain in winter months. Examples of all these techniques are described in the final section of the chapter.

The most obvious, and most common, way to use natural ventilation and thus avoid energy expensive ventilation and air conditioning systems is of course to open windows. It has the advantage that the degree of ventilation – and cooling – can be controlled directly by users. It has also the disadvantages that it can create drafts and, in buildings on heavily busy streets, bring in polluted air and produce unacceptable levels of noise. In large deeper plan buildings it is difficult to design so that the air circulates evenly throughout. One solution to providing fresh air throughout is to use the stack ventilation or chimney effect. In hot climates this has been known and used for hundreds of years – there are examples throughout North Africa and the Middle East. Air is brought in at the lower levels and rises by convection through the building; it can be filtered at the lower levels and channelled through all the working areas, providing a constant flow of clean air. One of the most extensive modern examples in Britain is the School of Engineering at Leicester De Montfort University, built in the early 1990s, which is also designed to use its thermal mass as a heat sink, levelling out fluctuations in temperature. The building was finished in 1993 and in 1997 it was still felt to be living up to its energy-saving claims – with one or two problems.[3]

Photovoltaic technology

This is relatively new and expensive – but is being improved steadily and thought by many to be capable of eventually making a major contribution to the efficient local use of solar gain. Greenpeace has estimated that up to 85% of the UK's energy needs could in principle be met by mounting photo-voltaic cells on buildings. That may be a little over optimistic and we are certainly a very long way from it at present.

The cells are electronic devices that directly convert the sun's light energy into electricity. They are mounted in roof panels and connected in series to batteries and inverters to provide useful power in the building. Unfortu-nately the systems so far used in Britain have proved to be highly un-economic. One estimate made for a proposed installation was that the system would take 73 years to pay for itself: the proposal was abandoned! A development in Ladbroke Grove in London for the Peabody Trust designed by CZWG Architects has been claimed to have the largest area yet of solar panels – 1500 m[2], but despite being heavily subsidised through European Union and DTI grants the costs to the trust are still estimated to be about 25% higher than a conventional roof.

However, as output increases costs are coming down. Figure 9.1 shows costs reduced and efficiency increased in the early years of photovoltaic cells, and the technology is developing all the time. The Canon head-quarters building in Reigate, Surrey uses what it claims to be a revolu-tionary new thin film photovoltaic technology, developed by Canon itself. The cells are less expensive to produce and are incorporated in a light-weight metal panel instead of the usual glass. There is a German system of roll-on roofing which incorporates photovoltaic cells and there will undoubtedly be many other experiments and innovations which should ultimately make photovoltaic systems economic for users as well as environmentally sound.

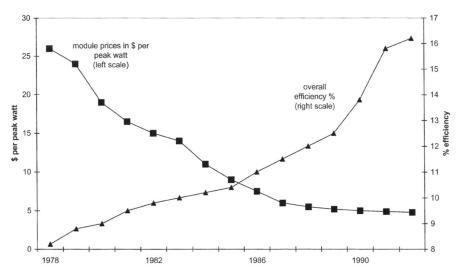

Fig. 9.1   Relative prices and efficiencies of photovoltaic cells 1978–1992.

Local wind generators

The use of wind power is becoming rapidly more widespread and more economic though still in the UK a very small contributor to total electricity consumption. However, there are now a few buildings – and more proposals – using small local generators to provide power for one, or a small group, of buildings. The Millennium Park Visitor Centre described below has a generator at the edge of its land. The architect Bill Dunster has designed a clover-leafed shaped residential tower which cleverly directs wind to drive a generator located in the central core of the building.[4]

Building materials and the environment

Some general considerations – sources of problems

So many thousands of different materials are used in construction that it is hardly surprising that many are now thought to be harmful in one way or another. It is probably impossible to envisage a system of construction that used none of these but their use can be limited or their damaging effects minimised. Contracts now often include 'blacklists' of materials which are not to be used – for many different reasons, not only potential environmental harm. However, these blacklists are not necessarily derived from sound evidence and they have caused much controversy in the industry. Several years ago Ove Arup and partners (as the practice was then called) together with the British Council for Offices produced a guide for specifiers *Good Practice in the Selection of Materials* to help identify materials which were not in fact harmful. The guide dealt with only 12 materials but there are more than 50 others that have been included in blacklists. Many of these too might be harmless but there is no doubt that the use of some materials *is* quite seriously damaging to the environment and that there is a need for a review of their use.

The harm that use of certain materials can cause takes many different forms. There are (two) major sources of damage:

● First the very process of obtaining the basic material may damage the environment from which it comes. Obvious examples are the quarrying of stone and slate and the extraction of clay; both can lay waste large areas of the countryside. Then there is the use of timber from threatened forests and species of trees; environmental groups such as Greenpeace have been engaged in a long battle to stop the destruction of large areas of tropical forests by excessive logging for timber much of which is used in construction.

● A second major effect is the pollution or environmental degradation caused by manufacture of construction materials. Major culprits here are cement, bricks and plastics, particularly PVC, all seen as basic inputs to the modern building process.

The next sections look at the two major examples of these damaging effects in more detail: first the impact of uncontrolled and destructive logging and secondly the polluting effects of PVC manufacture and use.

Wood, construction and sustainability

There is no doubt that wood is a sustainable resource; there is also no doubt that it has excellent properties as a construction material. It can be cut into virtually any shape; it has considerable compressive and tensile strength and with modern lamination techniques can be made into beams with the tensile strength of steel. It can be finished in many different ways to give aesthetically pleasing surfaces; it can be transformed into wide boards of many types and it has a high insulation value because of its cellular structure. Its major disadvantages – its vulnerability to rot and fire – can both be reduced, if not totally eliminated, by modern treatments. As the advertising campaign run by a prestigious group of timber associations

'Wood for Good' has claimed it is the greenest of all building materials as the forests themselves are 'the world's purifiers'.

There is a hint in that last phrase of where a problem lies; if the forests are the 'world's purifiers' their excessive destruction could be disastrous. The evidence from the scientific studies carried out over the past few years shows that despite the industry's claims and despite the fact that refor-estation is now carried out on a large scale, the rate of destruction of some types of forests and some types of tree, only a proportion of which, of course, are used for construction purposes, is creating damage which cannot be undone.

Photo (13)   Excessive logging is a threat not just to the tropical forests but also to temperate forests such as the Great Rainforest in Canada above. Copyright McAll-ister, Greenpeace, used with permission.

Something like 80% of the world's original forests have already been destroyed and of the 20% that remains nearly 40% is under threat mainly from large-scale logging to satisfy the demand for paper, construction and other uses. Many governments, non-governmental organisations and companies are now cooperating to reduce the rate of destruction of the remaining ancient forests, but in the face of rapidly growing demand the struggle is a hard one. An international organisation, the Forestry Stewardship Council (FSC) has introduced a set of forestry management standards and a labelling system which guarantees that labelled timber comes from properly managed sources. The aim is to reduce to its absolute minimum the use of timber from the clear cutting of the ancient forest – not only the threatened tropical forests of South America, Africa and the Far East but also the temperate and boreal forests of Canada, Northern Europe and Russia.

There has been considerable success; major consumers such as IKEA, B&Q and Meyer International (the UK's largest timber importer) have introduced purchasing policies to limit their use of wood from environmentally damaging sources. The effects have been dramatic with, for example, tropical timber use in the UK, Germany and The Netherlands dropping by 36% between 1992 and 1996.

The battle is not over, however; despite apparently strict controls in Brazil for example, even their own government organisations have admitted that up to 80% of logs cut in the Amazon forest are extracted illegally and that many of the forest management plans are in practice being ignored.

The part the construction industry can play is clear. It can ensure that the timber it uses comes from approved sources – that is, generally well managed sustainable forestry and not the clear cutting of ancient forest. It can reduce waste. It can use alternative, environmentally less harmful alternatives (not PVC! – see below) and it can use the most appropriate timber products for each specific purpose. An example of what can be done is the use by the Rugby Group of MDF panels for its doors which it reckons has reduced its use of ancient forest timbers by 60% each year.[5]

There is now plenty of guidance available to enable designers and specifiers to ensure that they can both use timber for all its excellent qualities and minimise the damaging environmental impact of its production.

The problem with PVC

The use of plastics in construction (as in virtually everything) has increased rapidly since the 1960s. It is only relatively recently that the serious environmental problems have been fully recognised. Of the many types of plastic, PVC, used widely for pipes, cables, windows frames and much else, appears to have the most damaging effects. Such is the concern that in some cities in Europe its use in some applications has been stopped and in Sweden the government has announced its intention of phasing out PVC completely. There are essentially two types of problem.

First, each stage of manufacture involves highly toxic chemicals, some of which are released into the atmosphere, and some captured in the waste products from the factories; these finish up in soil or in rivers and ultimately, the oceans. Some of these chemicals belong to the group called organochlorines, which can be highly persistent and build up in the food chain. The problem is made worse by the fact that many additives that are themselves toxic are used to make the basic PVC into a form usable for specific purposes. For example a chemical known as DEHP is added to make the material soft and flexible for applications such as cables. DEHP is a recognised pollutant and can leach from PVC during use. In the event of fire, uPVC produces highly toxic fumes, something which has been recognised by the use of alternative materials in electric cables.

Secondly, there are virtually insuperable problems of disposing safely of or recycling PVC products after they have been used. A detailed research study published in 1998[6] showed that as these have an average life of between 30 and 40 years we are approaching an enormous waste disposable problem. Although there are many recycling schemes, the processes involved are expensive, highly energy-intensive and are not in the end very effective. The recycled products are either remelted into a lower quality material – which postpones the waste problem but also can spread the toxic components further – or they are remelted with granulate incorporated into new product, so still involving more PVC production.

Other ways of disposing of used PVC products are incineration and landfill. Both are seriously polluting. In the 1970s incineration was carried out at sea but after scientific studies had demonstrated the impact on the marine eco-system the procedure was banned. There is considerable controversy over the damaging effects of incineration, which the industry claims only releases toxins in certain conditions. Landfill, the other main alternative, has its own problems – such as the leaching of chemicals into the surrounding earth and contamination of groundwater. These problems are not overcome even by deep disposal in old mine shafts for example – or as in Cheshire in old underground salt caverns.

There is still considerable debate about the seriousness of these issues. The PVC industry has argued in most countries that 'green campaigners' have exaggerated the problems – or even that there is no problem.

Obviously there is a great deal at stake, including the enormous investment in PVC production; but given the vast amount of research evidence now available, the case against seems overwhelming.

The alternatives

PVC products are now so ubiquitous in construction that it may seem impossible to imagine a building industry without them, but there are alternatives in virtually all applications. Many of these are the traditional materials that were used before good quality PVC components became available; they can offer comparable or even superior performance, need not be more expensive and with modern manufacturing techniques can be assembled efficiently. Other materials are 'modern', including other forms of plastic, but are thought to have a less damaging impact; one such is ethylene propylene rubber, used in cables. The Association for Environmentally Conscious Building has compiled a list of many such alternatives, which includes prices and manufacturer information.

The built environment as a whole

The most immediate environmental impact of construction is the physical change it brings about – in cities, towns and countryside. The construction industry has, after all, actually created the environment in which most of us live. Because of the virtual continuity of rural and urban areas there are difficulties in measuring the degree of urbanisation but it seems likely that at least 70% of Britain's population now live in cities, towns or their surrounding suburbs. Apart from the gardens, the parks and trees, most of our surrounding environment has been in one way or another constructed; and even the greenery has usually been planted as a part of the continuing urban development.

Cities were once seen as the heart of civilised living; in fact civilised meant to belong to a city (*civis* is the Latin for citizen). People moved to the cities or wintered in the cities not just for work but to escape the drudgery and deadness of rural life.

As the industrial revolution drew thousands more people in, the cities became places of appalling ugliness and misery (though retaining as did London, Manchester and Liverpool their pockets of upper class elegance). As public health legislation began to have an effect at the end of the last century (more is said about this in Chapter 10), things did begin to improve. Yet it was still possible for Thomas Sharp to write in 1940[7]:

'Our towns have been repulsive and inefficient for a hundred years or more – for so long in fact that most of us have become inured to their badness. Even those who are still aware of the possibility of a town's being an orderly, even beautiful, creation have come to regard the building of fine towns as a faculty which has somehow been denied to Englishmen.'

The same might well be said today.

Still in Britain and Europe today there are vast areas of dreary urban and suburban development which many of its inhabitants find depressing and constraining rather than a joy to live in – though human tolerance of bad physical conditions is high when compensated by lively community life. Yet the odd thing is that though we seem content to live in places of extraordinary drabness we do appreciate beautiful towns and cities elsewhere. We go to visit them in vast numbers – so much so that places like Venice and Florence are overwhelmed by tourists. It just seems that apart from a small minority – mainly Londoners – we no longer see our own urban environment as a source of pleasure.

Instead we move to the countryside. That seemed to the Victorians (those who could afford it) the only way to escape the dirt and noise of the city, and it is an attitude that survives. Today, urban dwellers in their thousands like to visit the countryside for recreation, for the landscape, the fresh air and the open space. The problem is that in doing so they are often helping to destroy the very qualities they are hoping to find – a problem which is particularly acute in a small highly urbanised country like Britain.

There are then two connected environmental issues here, both of which involve the activities of the construction industry. There is the issue of quality in urban areas, of which a dominating aspect is now the use of the car as a means of transport, i.e. of improving the urban environment; and the issue of the destruction of our remaining countryside. Governments (of all political persuasions) tried to tackle both of these problems for most of the twentieth century (though as we will see in the next chapter there were isolated attempts to control building right back to mediaeval times); yet pressures of population, industrial need and most of all the development of the internal combustion engine seem to have undermined much of what was attempted. There were major advances mostly through the town and country planning legislation and its associated system of planning control. For most of the time up to the beginning of the twentieth century there was very little restriction on where people could build. If you wanted to build a factory in the Yorkshire Dales or dig a mine in the Welsh valleys – or for that matter build a canning factory in California – you acquired the land by fair means or foul and carried out the work needed for your enterprise.

But increasingly towards the end of the nineteenth century while the new urban centres were still expanding rapidly, many people saw the dangers and proposed ways of dealing with them – people like Ebenezer Howard. His ideas for new garden cities and the logical dispersal of people into complexes of mutually supportive and self sufficient areas of town and country were first published in 1898 and then re-published with the title *The Garden Cities of Tomorrow* in 1902; in some ways it has been the most influential book on planning ever published.[7]

The first serious attempts at planning control were made in the 1909 Planning Act but it was a very weak affair and though it was strengthened by further acts of parliament such as the act of 1935 designed to control 'ribbon development' – that is, the growth of long arms of development from the cities clustered round the railways and later the roads – there was still no really effective planning system.

The debate actually intensified during the Second World War as part of the creation of a new vision for a new beginning that would come, it was hoped, with victory. Abercrombie's County of London Plan, a model for all future plans, was published in 1943. It was however the great Planning Act of 1947 that produced what has been described as the most comprehensive planning system in the world. The act itself gave very wide powers to local authorities both to control private development and to develop themselves; it was followed by further important policies which had enormous impact – the New Towns Act, the creation of the national parks, and the creation of green belts around cities.

Yet despite the successes, the pressure on the countryside continued and still continues; green belts are broached, national parks struggle to contain development and new suburbs spread inexorably. Despite the very wide and, some builders would argue, draconian planning powers that local authorities have, urban development did not become notably more humane. In fact it has been argued that local authority planners were themselves guilty of conniving with the construction industry to produce some of the worst environments that we now have in some high-rise estates and soulless suburbs. In fact all the problems are still with us, though it does seem that recent attempts by governments to tackle them may bear some fruit.

For example, there is now considerable pressure on developers and builders to reuse urban land – the so-called brownfield sites. The 1995 White Paper on Rural England set a target that half of all new homes should be built on such brownfield sites by 2005 but in fact by 1993 already 49% were being built on previously used vacant land – an increase from 11% in 1988. Most of that was done by the large housebuilders without government pressure, as usable sites became available in areas where demand was high.

It is a controversial subject – where housebuilders have clashed with government on the actual necessity and extent of such control. When new plans were announced in 1999 to give local authorities further powers to restrict development on green land, the housebuilders were reported to be vehemently opposed. They argued that brownfield land is usually expensive to build on and as often as not is in areas where there is little demand; for example it was said that there were plenty of brownfield sites in the Merseyside and Manchester areas – but people wanted to move out to the more rural areas of Cheshire, which illustrates the fundamental problem we have.

Countryside preservation, urban containment and urban design are all vast and complex issues; the only point of touching on them so superficially has been to indicate their importance. Environmental improvement in building means more than just better insulation or photovoltaic cells, it is at the heart of the way we live.

Where we are now

The encouraging aspect of the situation today is that in all of the areas discussed above, the efficient use of energy, the use of environmentally friendly materials and the impact of building on the whole environment, there *is* a growing awareness of the problems within the industry, with more attempts to solve them and many successes. The stream of research reports, academic analyses, government documents and design guides has become a fast flowing river. It is difficult to keep up with it all – particularly if your immediate concern is getting the next contract. Yet the outlines of what needs to be done are fairly clear.

There are now many examples of buildings and developments which have been designed to minimise harmful environmental impacts and many of these have taken into account all the factors discussed in this chapter – and indeed some that have not been discussed.

The Millennium Village scheme at Greenwich, already referred to above, is designed as an exemplary energy-saving project. It is intended that it will use 80% less energy than conventional housing, half of the saving coming from reducing energy use and half by using efficient forms of energy supply and conversion. The high levels of glazing on the south-facing facades will increase passive solar heat gain; a combined heat and power plant will supply locally produced electricity and reclaim waste heat. High levels of insulation will reduce heat losses as will the use of wall cladding with pre-fitted window frames to reduce air leaks. Other 'green' features include the use of bathwater for flushing toilets and the use of low energy construction materials.

The Millennium Village is a purely private scheme, but some housing associations have been particularly enthusiastic in moving towards environmentally sensitive developments. The Peabody Trust has also taken a lead in factory-produced housing as described in Chapter 8, and has produced some of the most ambitious low-energy schemes in the housing field. Its Beddington zero energy development, BedZed for short, designed by architect Bill Dunster, incorporates an impressive list of energy-saving features including all those discussed earlier in the chapter plus some more. They include (among others):

- it is built on 'brownfield' land – a former sewerage works in fact
- it is a high-density scheme, using less land than a conventional housing estate of that size – but it appears spacious; all houses have back gardens, balconies and roof gardens
- dwellings include office space – so people can work from home
- an electric car pool will make local personal transport available without the need for everyone to buy their own cars
- the combined heat and power plant uses waste wood as fuel
- materials used in construction were as far as possible locally sourced and recycled
- buildings (in five rows) are oriented and glazed to take maximum advantage of solar energy.

Outside housing there have been many commercial and industrial and leisure developments in which designers have aimed at minimum energy use; some of these, such as the School of Engineering building at Leicester and the new Canon factory, have already been referred to. Others include the National Energy Centre at Milton Keynes (if they cannot do it who can?). The building uses natural ventilation and is highly insulated. It maximises natural lighting by using reflective light shelves on the north side and reduces unwanted solar gain with external metal louvres as well as internal glare control blinds.[8] The same principles are used in what the Building Research Establishment called the most energy-efficient building ever, the Wessex Water headquarters near Bath.[9] The Millennium Park Visitor Centre near Leicester is another demonstration project which uses almost every available device – solar panels, optimisation of solar heat and light through careful orientation and window design, high insulation, recycled materials, stack ventilation for cooling and electricity generated from a local wind turbine; rainwater supplies all the centre's water requirements, which are minimised by the use of low flush and composting toilets.[10]

There are now dozens of examples which together are beginning to demonstrate that the objective of reducing the impact of construction on the environment through reduction of energy use and all the other ways discussed here can be met, given the will and the determination. It may still need a little more help from government – which raises for the industry a spectre we shall look at in the next chapter.

For further study

The best immediate source of information on sustainable building is the Building Research Establishment and its specialist energy division BRESCU. The BRE website www.bre.gov.uk leads to a vast amount of information. The publications list, which is downloadable, is 83 pages long! Publications include case-studies, research reports, advice and guidance.

Greenpeace International, often thought of as simply a campaigning organisation, has sponsored and produced some excellent scientific research. The website is www.greenpeace.org

The Centre for Alternative Technology at Machynlleth in Wales produces advice on a range of the issues discussed in this chapter and can provide catalogues of information, research and case-studies. The website is www.cat.org.uk. Their physical site is worth a visit as well!

There are many books in the field; an interesting and comprehensive one on housing is *Sustainable Housing: Principles and Practice* edited by Brian Edwards and David Turrent, published by E. & F.N. Spon in 2000.

Roaf and Hancocks' *Energy Efficient Building* was published in 1992 but is still interesting and relevant. Peter Graham's *Building Ecology* gives a broader view of the relationship between building and the natural environment. Adams and Watkins *Greenfields, Brownfields and Housing Development* looks in much more detail at the issue just touched on in this chapter.

10 The Government and the Building Industry

- Introduction: the state and construction – control and support
- Ten points of contact
- Case study 1: the Building Regulations
- Case study 2: cowboys

Introduction

In every chapter of this book some reference has been made to government impact on the construction industry – the government and its agents (collectively, 'the state') as client, as policy maker, as monitor, inspector, controller, regulator but also as supplier of information, source of finance and in many ways *supporter* of the industry. Although each of these has been referred to in different contexts, we can, by bringing them all together in this final chapter, underline just how all pervasive the state's influence is – as well as providing a brief summary of much of the earlier discussion.

The chapter is structured as follows: the points of contact between government and construction which have been described earlier are listed – with some comments on their implications. This initial listing is followed by two 'case studies' which illustrate rather different forms of intervention and between them point to some possible answers to the controversial question which is continually raised by and about the industry: is there too much or too little intervention by the state?

The first case study is concerned with the building control system – which is deeply embedded, has a long history and is by most standards, vigorously enforced. The second section is a brief history of recent attempts to do something about so-called cowboy builders. Here we will see rather desperate attempts to use government regulation as a means to control what may not be controllable; it may indeed illustrate the limits of what a government, with the best intentions, can achieve. Finally and briefly we look at some of the controversial issues raised by all the examples. Does government at national or local level interfere unnecessarily with the industry's operation? Would society benefit more if the construction market were just allowed to operate untrammelled? Or is the opposite the case – that there is still not enough regulation to ensure that society as a whole gains maximum benefit? These are highly political questions and we will ultimately leave the answers to the reader.

Industry and state – a review of the points from earlier chapters

The state as monitor

Chapter one referred to many of the reports that have been produced over the last two centuries. In the late eighteenth and early nineteenth centuries, the motivation was usually to ensure that the government's own buildings were being provided in the most economical way. Later, as concerns grew about housing conditions, government accepted responsibility for ensuring reasonable standards. When, after the First World War, state subsidised housing was produced on a large scale, there were frequent investigations as to whether the industry could cope with the demands made on it. After the Second World War and throughout the second half of the twentieth century, the government took on a much broader remit; construction of infrastructure and of industrial buildings, the rebuilding of city centres, the new towns and the new council estates all required an industry that was efficient; there was a stream of reports examining many different aspects of efficiency. Sometimes their recommendations were acted on, sometimes they seem to have been ignored. As was suggested in Chapter 1 the Egan report can be seen as the end (so far) of a very long line.

Apart from the major official reports there has been and still is a constant stream of other documents reporting and commenting on all aspects of the industry. The bi-monthly *Construction Monitor* now produced by the DTI, reports developments in research, important conferences, new regulations and guidelines and summary statistics.

The state as client

As the discussion and statistics in Chapter 2 demonstrated, the state is itself a major client for the industry and if thought of as one collective unit (which makes some sense as the finance is centrally determined) is the biggest single client. It inevitably has, therefore, a potentially dominating impact on the industry – and some of the implications were examined in Chapters 3 and 8.

There are two different aspects to the impact of the state's power as client.

First it actually determines the level and type of demand which the industry has to meet. The amount of roadbuilding work available, for example, depends entirely on political decisions. This is true for the vast majority of hospitals and schools – *even if they are procured as PFI projects.*

Secondly its position as major client has enabled it to promote new techniques of procurement and construction such as partnerships and prime contracting, low-energy design and use of least environmentally damaging materials. As we saw the public–private partnerships and PFI have forced the industry into thinking of its work for the public sector as provision of a whole package of services over time – not just the construction of a building or civil engineering project.

The state as economic policy maker

The arsenal of policy weapons that governments use to promote objectives such as economic growth, full employment, low inflation, regional development, urban regeneration and so on is vast, and virtually all of them have a powerful and immediate effect on the industry. Some have been discussed earlier – such as interest rate policy, special funding initiatives and youth employment schemes. Others have hardly been mentioned but may be equally significant. For example, the long series of urban regeneration initiatives (actually stretching back to the 1930s) have provided finance and stimulation for a great deal of construction work. The enterprise zones of the early 1980s, followed by the urban development corporations, poured millions of pounds of government finance and levered millions more from the private sector into areas such has the docklands in London and Liverpool. Between 1981 and 1988 over £300 million of government funding and over one and a half billion pounds of private funds flowed into the London Docklands area. From that initial stimulation followed the massive development of the area, beginning with Canary Wharf, which is still going on. Similarly the Merseyside Docklands Corporation initiated a process of development that has transformed miles of derelict dockland into a busy commercial, residential and leisure area.

Other areas of government policy which have at first no obvious construction implications often turn out to have some requirement for new building. The creation of a freeport in Merseyside for example – part of government policy to increase foreign trade – has similarly led to the construction of new factories, storage units, warehouses and offices.

The state as legislator

Parliament makes laws covering every aspect of our lives; some are major pieces of legislation to meet major objectives of government policy, many others are specific laws aimed at resolving particular problems within particular sectors of society and industry. Both have affected construction. Many examples have already been described: the Town and Country Planning Acts in the last chapter, the Construction Act in Chapter 7, the Health and Safety at Work Act and of course all the many housing-related acts which have reached the statute book every few years since 1919. The acts of 1969 and 1974, for example, opened up opportunities for the development of the housing associations. These have since grown, with the help of lots more legislation from a peripheral group of small housing charities to the major provider of so-called social housing (described in Chapter 2).

The state as collector of taxes

This could be considered as part of the state's economic policy but it is more than that. Taxation existed well before states had economic policies, when kings taxed to fund their wars and their wives, not their policies. Taxation affects every aspect of the construction firm's work. Chapter 4 examined one particularly significant area – the tax status of self-employed workers. Value added tax is another – particularly controversial – form of taxation for the industry because of the fact that new construction is VAT-exempt but repairs and maintenance are not. It has been suggested that redevelopment on brownfield sites discussed in Chapter 9 could be assisted by adjustments to the VAT regime.

The state as regulator

As well as acts of Parliament, government has powers (established by particular acts) to make regulations. For example, the Building Regulations can be altered without Parliament having to go through the complex procedure of drafting and passing a new act. Changes are not just made arbitrarily, however; there is a long process of consultation and the seeking of expert advice before changes can be 'laid before Parliament' and agreed. The construction industry tax scheme already mentioned was introduced as a set of regulations under the general powers given to the Inland Revenue. The CDM regulations (described in Chapter 4) were brought in under the powers of the Health and Safety at Work Act. A more extended examination of the Building Regulations themselves is included below as one of the two case studies.

The state as provider of finance

As a major client, the government obviously puts money into the industry through its procurement of projects. But there are other channels through which the industry has access directly or indirectly to public finance including the multitude of subsidies and grants for various forms of private construction. Financial assistance of this sort has been introduced since the end of the First World War in all sorts of different contexts. Local authority housebuilding was subsidised throughout its history, there was subsidy for a short while for private housebuilders, there have been subsidies for business relocation, for factory units, for farm buildings and many others. Important examples today are the Housing Corporation Grant which still supports most housing association (registered social landlord) schemes, the Derelict Land Grant making development on reused sites more economic and the many forms of grants and subsidy to encourage urban renewal.

The state as source (and co-ordinator) of information

Much of the information used in this book has come from government sources – particularly the National Office for Statistics which provides the most easily accessible and comprehensive figures on construction and housing. Apart from the specific construction statistics there is a mass of other relevant data – for example, on the economy as a whole, including forecasts of future activity, analyses of costs, wages, incomes, population and household structures.

There is also a large output of information on all the things discussed above – from the Inland Revenue, for example, on all forms of taxation, from various departments on new laws and regulations in every field, and then there is the growing mountain of documents on European Union regulations and their implications for British practice.

The state as research sponsor and technical adviser

The Building Research Establishment is now theoretically independent but was, as the Building Research Station, a government body responsible for carrying out and sponsoring technical research into all aspects of construction. It has produced hundreds of publications which describe research results and give authoritative advice on, for example, the use of particular materials.

The government is also a major sponsor of construction research in universities and other institutions and disseminates the results of the most significant work through its various publications.

The state as facilitator

This sounds a little vague but the government's action in bringing people in the industry together – with each other and with people in other industries – has been an important factor in the development of many construction organisations. For example, after recommendations in the Latham report, the Construction Industry Board was set up. It was wound down in June 2001 but was replaced by the Strategic Forum for Construction – to be chaired by Sir John Egan.

Other aspects

There are many other points of contact and influence between the state and the industry; it is often said that construction is the most regulated of all industries, a point difficult to prove but reasonable to believe. Instead, however, of embarking on what might turn out to be an apparently infinite list, the chapter – and book – ends with a slightly more detailed discussion of two areas of control which in most ways are in total contrast to each other and yet both raise the same question – too much or too little?

Case study one – the Building Regulations

Origins

Probably the most extensive detailed prescriptive and rigorously enforced mechanism through which the state exercises influence over building, and certainly the one with the longest history, is the building control system – that is, the Building Regulations and their implementation. The requirement for buildings to comply with the regulations is such an integral part of the building process that it is now taken for granted. It is still, however, capable of generating plenty of controversy, especially when amendments and additions are proposed and introduced as they frequently are. Strangely, in many discussions of the relationship between government and the industry, the Building Regulations are hardly mentioned or treated as something purely technical. Yet they often are and always have been highly politically contentious and they have changed to reflect political priorities. When the industry's representatives have complained of being over-regulated it is usually some element of the building control or planning system that has sparked off their anger.

The fact is that all industrialised countries and, in fact, most countries in the world have similar regulations; they may be less extensive elsewhere and in many countries are not as thoroughly implemented as in others, but the need for control is universally recognised. In fact the imposition of requirements on owners and builders goes back a long way and diligent historians have found examples from over 3000 years ago. They need hardly concern us here but the origins and history of the British regulations are interesting and relevant in that at each stage of their long development they have retained elements of previous forms of control.

It was the city of London which imposed the earliest regulations; it was also London which developed the most detailed and extensive system of its own, retaining its independence and individual requirements, even after the national system was established, right through to 1984. Even as early as the end of the twelfth century, London had begun to establish rules which seem mainly to have been for three purposes: prevention of disputes between neighbours, prevention of fire and prevention of encroachment of buildings over roads.

There were detailed provisions for the thickness and height of party walls and a right to require a neighbour building a house not to block off light (hence 'ancient lights'). Methods to reduce fire risk go back well before the Great Fire of London – citizens being required in the thirteenth century to roof their houses in lead, tiles or stone. A curious example of rules about encroachments on the highway was the regulation that any projections built at first-floor level should be high enough to allow a person on a horse to pass under without banging his head.

Most of these early rules were ignored, as there was no system for monitoring or controlling compliance. The same happened to the many attempts to control the scale of building in London in the Tudor period, when regulations were made to prevent any new unauthorised construction in London itself and to control densities throughout the country. Some of the rules were obviously impossible to impose such as the one requiring – throughout the country – that no-one should build a house with less than four acres of ground. A proclamation of 1602 complained, in regard to attempts to control speculative building, that

'Notwithstanding her (i.e. Queen Elizabeth's) gracious and Princely commandment ... yet it has fallen out, partly by the covetous and insatiable dispositions of some persons, that without respect to the common good and public profit of the realm, do only regard their own particular lucre and gain and partly by the negligence and corruption of others who ought by reason of their offices to see the proclamation performed do undutifully neglect the same ...'[1]

Despite the appalling punishments to which lawbreakers were liable in those days, the rules were still ignored; it was to be a long time before it was fully appreciated that proper systems of monitoring and enforcement were necessary, if regulation was to be effective.

All these early attempts were spasmodic, not very coherent and largely ineffective but a series of proclamations under King James and his successor Charles I from 1619 onwards established real precedents for the modern building acts. They imposed extremely specific requirements on builders and owners; for example

'... and if the said building do not exceed two stories in height then the walls thereof shall bee the thickness of one bricke and half a brickes length from the ground to the uppermost part of the said walles ... in the building of the said houses there shall be no jutting ies or jutting or Cant windows either upon timber joists or otherwise but the walles go direct and straight upwards ...'

(a final solution to the problem of horse riders banging their heads?).

It is interesting that some of these regulations were included for aesthetic as well as structural and fire prevention reasons:

'... All shops in every principal street of trade be made of pilasters of hard stone or brick and the heads of shop windows cut in wedges Archwise to sustain the wall about it and *for the ornament of the streets'* (emphasis added).

It was around this period that detailed specifications and standard sizes were laid down for bricks to be used in London; and there was even an attempt to fix the price. The clay to be used to be 'good and fit for its purpose' dug only at specific times, the final size was to be $9'' \times 4\frac{1}{8}'' \times 2\frac{1}{4}''$; and the price eight shillings per thousand at the kiln.

Although all these regulations were detailed and comprehensive, it required, as so often, a major disaster to stimulate a major step forward. The fire of London of 1666 destroyed four fifths of the city. Most of its major buildings were lost – including St Paul's, the Guildhall, the halls of the City companies, 87 parish churches as well as over 13 000 houses. The story of the rebuilding of the city is complex, but its relevance for our purpose here is that it produced the first real Building Act which served as a model for all that followed; in fact the historians Knowles and Pitt claimed that it was far superior to most of its early successors.

It contained 48 detailed sections. They defined the only four types of buildings that were to be built and specified in great detail how each was to be constructed. They were to be built of stone or brick (or both) with timber used only for window and door frames. There were rules relating to party walls, to foundations and to roofs. One innovation that became a standard part of future regulations was the use of tables and diagrams to clarify their exact meaning.

Over the next 150 years further acts, including a major one of 1774, further refined and extended the regulations, including many which were to remain a permanent part of the control system in London – but also many which disappeared – such as the duty laid upon householders to 'sweep and scrape the pavement in front of their houses before 10 o'clock every morning'. One important advance was the setting up of the posts of District Surveyors – which remained unique to London – defining at last a clear responsibility for ensuring regulations were complied with.

Public health and public responsibility

Most of the building acts and associated regulations were chiefly aimed at preventing the spread of fire and ensuring structural stability. Similar but less extensive regulations developed in other cities but there was no national system of control. Indeed even the old controls were weakened in the new political climate of *laissez-faire* – the belief that government had no right to interfere in the working of the free market and the economic decisions of individuals.

This all began to change in the middle of the nineteenth century when the full horror of uncontrolled development in the new manufacturing centres and ports eventually began to disturb the conscience of at least some of the city and national politicians. There was in fact fierce political struggle – lasting most of the century and at its most intense between 1840 and 1875 – to establish that governments not only had the *right* but also had the *duty* to impose controls. Individual cities brought bills to Parliament to give them powers to regulate the density and types of dwellings allowed; but they were always fiercely opposed – even by the cities' own representatives. Some of the proposals indicate just how bad conditions were. A bill brought forward by Lord Normanton in 1841 included a clause to ban cellar *dwellings unless there was an open space in front of at least 3 ft*. It was defeated.

The breakthrough came with the Public Health Act of 1848, not because the country's rulers had become soft hearted, but because they feared the spread of cholera and typhoid from the dense working class areas to the more salubrious parts, and it was bitterly opposed. *The Times* and *The Economist* wrote angry editorials against such outrageous interference with the right of free citizens to build what they liked. One example of the outrageous things that were being proposed was a duty of local authorities to provide sewers; in fact the act weakened 'duty' to 'right'.

The Act gave local authorities powers to establish by-laws to control building and standards in their area, powers which most of the big cities took up with more or less enthusiasm. The 1848 Act was reinforced and partly replaced by the great Public Health Act of 1875 which is now seen as the real progenitor of modern national buildings regulation systems.

There was now a new dimension to regulation; it was imposed not just to minimise the risk from fire and building collapse but to prevent diseases and the spread of disease; and simply to make cities more humane places to live in. This was not actually the first time that such motives lay behind regulation – the early London acts were often attempts to prevent plague as much as fire – but it did represent a wholly new direction in the nature and purpose of control.

Towards comprehensive regulation

Later acts at the end of the nineteenth century and through into the twentieth century consolidated and further developed the principles established in these early public health acts. Changes have been implemented usually by the passing of an enabling act at intervals of ten to 20 years, giving Parliament the powers to amend the details of the regulations themselves. The developments were of three sorts:

- First, more aspects of construction were subject to control as new techniques, new materials and new priorities were recognised. Conservation of fuel and power, for example, is a relatively recent addition; insulation standards were first introduced in 1965. Regulations relating to facilities and access for disabled people (part M of the current regulations) are even more recent, first introduced (then as Part T) in 1985.
- Secondly, the methods of enforcement were extended. In an attempt to bring in a greater degree of self-regulation the 1984 Act introduced a system of private certification – which involves using non-local authority, but formally approved, inspectors.
- Thirdly, the criteria on which compliance was defined have been steadily broadened. The old form of regulations were specific – they laid down exactly what was to be done – in terms, for example, of the sort of material to be used, the exact measurements of beam sizes or distances between elements. These have been largely replaced by more flexible functional and performance standards. That is, compliance is demonstrated by showing that an element or part of a building performs its defined function or meets a required level of performance. For example the U-values discussed in the last chapter define the maximum heat loss allowed through, say, a wall; they do not say how the wall is to be constructed. The advantage of this sort of regulation is that it gives designers and builders much more flexibility and allows for innovation. The requirements are now most frequently expressed as requirements to comply with specific British Standards and Codes of Practice.

The current Building Regulations derive their statutory force from the Building Act of 1984, the most recent fully amended set being those of 1991 (since when several detailed amendments have been issued). The regulations themselves are concerned with definitions and implementation. What are normally referred to as the regulations are actually the *requirements* of Schedule 1 of the regulations; these are extended in a standard set of *Approved Documents* which give detailed guidance on how the technical requirements of the regulations can be met. Figure 10.1 lists the subjects of the 13 parts of the current requirements and approved documents.

This list gives an immediate indication of the range of control as it exists today; the comprehensive coverage and the requirement that most building work (there are defined exceptions) has to be submitted for approval before commencement and can be checked at any stage means the control is very strict. It also means that building control inspectors, whether local authority

PART A: Structure – loading, ground movement and disproportionate collapse
PART B: Fire safety – means of escape, spread of fire, fire service facilities
PART C: Site preparation and resistance to moisture – preparation of site, dangerous
  substances, resistance to weather and ground moisture
PART D: Toxic substances
PART E: Resistance to sound
PART F: Ventilation – means of ventilation and condensation in roofs
PART G: Hygiene – bathrooms, sanitary and washing facilities, hot water storage
PART H: Drainage and waste disposal – foul water drainage, cesspools, septic tanks, etc.
PART J: Heating appliances
PART K: Protection from falling, collision and impact – stairs, ladders, ramps, protection
  from falling, from impact with windows and doors
PART L: Conservation of fuel and power
PART M: Access and facilities for disabled people
PART N: Glazing – safety in relation to impact, opening and cleaning

Fig. 10.1   Structure of the Building Regulations (Schedule 1 requirements).

employees or, under the 1984 system, approved private inspectors, are sometimes not the most popular people with builders.

However the point of relating the origins and growth of the system above was to show just why it has arisen; regulation was a response to real abuses and real dangers. There will always be arguments as to whether they are too prescriptive, too detailed or too comprehensive; it will always be possible to pick the odd anomaly or absurdity – particularly of building inspectors' decisions. Building Regulations can sometimes be the wrong weapon to use: at the beginning of the nineteenth century a regulation required wider chimneys so that sweeps should not get stuck. It was some time before it was decided that a better way was to ban sending small boys up chimneys. Yet we should not lose sight of the fact that the regulations have given us buildings that are safe to live and work in. The occasional failures have occurred either because features of the building were not covered by regulations or regulations had been ignored.

Objectors to this argument in the industry say that we would have safe building without all this regulation and to a degree this true; but partly because the standards have been established as normal practice, often through the regulations themselves. History and the experience in other countries show that weak regulation and weak control can have disastrous effects.

The story of the 1999 Turkish earthquake is salutary but also raises other questions. Many buildings collapsed; many did not. The ones that did had not met the current regulations. For example, some low structures with light foundations had had heavier storeys built over them; in some the wrong sort of steel had been used or there was insufficient steel reinforcement. The head of Ove Arup's earthquake engineering department was quoted as saying:

> 'Turkish building codes are among the best in the world; the weak link was the implementation and enforcement of the regulations and the responsibility for that lies with the municipal authorities ... we shouldn't blame the contractors; they are going to build to minimum standards ... it's the nature of the game.'[2]

That is a debatable point but just to complicate the issue and perhaps provide further food for thought, it should be added that among the buildings that did *not* collapse and were *not* damaged were a number of ancient mosques – built well before building regulations were heard of.

Case study two: cowboys?

Every few years the problem of so called 'cowboy builders' surfaces as a major issue – something that needs to be dealt with urgently. Why it is seen as more important at one time than another seems to depend on the publicity given to clients' problems in the press and on television. In the late 1990s a number of popular TV programmes from *Auf Wiedersehen Pet* to one called '*House of Horrors*', which portrayed small builders and building workers as everything from lovable rogues to outright crooks, alarmed the industry for the image they were spreading but which were recognised as familiar reality by thousands of people who had suffered. The fact is that the industry – like every other industry but perhaps more so – has always had its share of unscrupulous and/or incompetent builders. The problem was probably worse in the past than today – particularly before the building control system was fully developed. As mentioned in Chapter 1, the term jerry builder became commonly used in Victorian times to describe many of the people who were throwing up houses at great speed to meet the huge new demand as cities grew. But there has been no period (and it is the same or worse in other countries) when the problem has *not* existed.

What has changed more recently is the general sensitivity to consumer concerns, consumer groups and the growth of consumer protection – through legislation and the many different consumer organisations. For example, in 1996 the National Consumer Council produced a report called *Controlling the Cowboys* calling for a national database on builders and an official code of practice. As well as anecdotal accounts in the press, statistics showing the scale of the problem were widely publicised; in 1996 the Office of Fair Trading reported it had received 93 000 complaints about home maintenance – more than complaints about secondhand cars.

Various attempts at effective action have been made. In 1988 the government published a report *Beating the Cowboys* but ten years later another government report said, 'there is little evidence that it had any lasting effect'.

Some proposals and some action came from outside the industry altogether, the Automobile Association setting up an Approved Supplier scheme for example. The industry itself became increasingly alarmed and came up with various proposals – such as an extension of the Chartered Building scheme (this seemed a bit optimistic as the scheme had only 300 members at the time). The proposal which seemed most attractive to the larger builders was a reduction in the rate of VAT, which smaller firms could avoid, enabling them to undercut those which were VAT registered. Interestingly, in view of the usual assumption that small jobs are not economic for large firms, Mowlem Construction set up their Skillbase scheme 'to provide a 24 hour domestic maintenance service direct to the UK's 16 million home owners' which an article in *Building* even suggested could 'drive the cowboys out of town.'[3]

Renewed government action came in 1998 when the Labour government set up a task force under Tony Merricks, manager of Balfour Beatty's specialist contracting business, to look at the whole issue and make recommendations. After much taking of evidence, much discussion and not a little acrimony, the task force came up with an interim report in April 1999 which was put out for consultation and a final report five months later.

The reports identified three types of 'cowboys':

● the incompetent builder, who lacked the skills and knowledge to do work he had undertaken, sometimes even unaware of his duties under the Building Regulations
● the dishonest builder, deliberately defrauding people such as the elderly and vulnerable
● the tax evaders, failing to pay income tax or VAT; these not only deprived the government of revenue but were competing unfairly against other legitimate and tax-paying contractors.

The report's main recommendations were that there should be a nationwide register of builders who had achieved a 'quality mark' and that their work should be backed by a warranty scheme. By the end of 1999, Birmingham and Somerset had been chosen to pilot the proposed scheme and a technical panel was set up to develop the detailed criteria for the quality mark.

Although the report was welcomed – indeed highly praised – by people on all sides of the industry, by politicians of all parties and by consumer groups, many were disappointed. The main objection – and one which may prove to be well founded – was that the proposed action included no statutory obligations. Builders would have to be persuaded to join the scheme (which would actually cost them money), the incentive being that they would have a recognised quality status and would be on a register. Anyone requiring building work with some guarantee of quality would be able to find their names and be expected to prefer them to unregistered builders.

The major problem, as many commentators pointed out, was that as so many consumers will prefer to have work done at minimal cost, they can still be taken in by the unscrupulous or incompetent builder. Large-scale publicity, warning the public to use only 'quality mark' builders might make a difference, but deeply embedded black markets are not easily defeated. This was made very clear by a survey carried out by the National Federation of Housebuilders which showed over 50% of people would opt for a builder who did not charge VAT.[4]

Scottish Building, a body representing 1400 companies, is in fact calling for a mandatory scheme; small firms would be required to register and submit to tests of competence. Before the 2001 election, the Minister for Construction had also said he would consider compulsion if the voluntary scheme was not effective.

It is not at all clear how such a scheme would work but the intriguing thing is that we have members of an industry that has complained so often and so bitterly about government interference and over-regulation actually asking for more control.

Conclusion

Clearly there is some ambivalence in the industry's attitude to the state and the state's attitude to the industry. But it is an indication of the social importance of construction that government's intimate involvement in the construction industry is almost certainly unavoidable and it is probably in general a good thing. The recent change in responsibilities with the reorganisation of government departments after the 2001 election has been

greeted with typically contradictory reactions, but with some strong feel-
ings.

The issues raised in this chapter are fundamental ones not only with
regard to construction but to all industry. The right balance to be struck
between profitability and social responsibility is not an easy one to find but
the continuing tension between state and industry may nudge us towards
some reasonably satisfactory solution. The Building Regulations are
possibly an example of where that balance has been struck; in the case of
'the cowboys' there is still some way to go.

The debate will continue about what kind of control there should be and
where the boundaries are to be drawn.

Endpiece

This book has examined at an introductory level many aspects of the construction industry – its markets, its structure, its workforce, its modes of operation and some selected elements of the social and political context in which it works. There is much that has not been included but might have been – the details of management on site, the relationship of construction firms to the materials industry and the increasingly significant international character of construction, for example.

Hopefully there is enough here not only to give some understanding of the industry but also to provoke some discussion and further investigation into the many problematic issues the industry presents. It is possible to read the daily newspapers, listen to or watch daily news broadcasts and have no idea what is happening in the world of construction – unless some disaster strikes. Yet the industry's work is vital to our everyday existence; unless we are on a wilderness expedition, we cannot move without being influenced by the surrounding built environment. It affects us physically and aesthetically; it determines where we work and live and in what conditions.

The industry's public image seems on the whole to be pretty negative; but when some really exciting project is successfully completed – the Eden Project in Cornwall, the Lowry Museum in Salford and the new Tate Modern are all recent examples – people react with enthusiasm. The design and construction of buildings has been and can be one of society's most exciting activities; the construction industry, with its enormous responsibility of defining and creating our built environment should be and can be one of the most exciting and interesting in which to work.

Notes and References

Chapter 1

[1] Sir Michael Latham (1994) *Constructing the team: final report of the government/ industry review of procurement and contractual arrangements in the UK construction industry*, HMSO, London, July, referred to as the Latham Report.
Department for Environment, Transport and the Regions (DETR) (1998) *Rethinking Construction: the report of the Construction Task Force to the Deputy Prime Minister, John Prescott, on the scope for improving the quality and efficiency of UK construction* The Stationery Office, London, referred to as The Egan Report or *Rethinking Construction*.
[2] *Modernising Construction* (2001) Report by the Comptroller and Auditor General.
[3] Marian Bowley (1966) *The British building industry – four studies in response and resistance to change*, Cambridge University Press.
[4] *Training for the building industry* (1943) Cmd 6428.
[5] *Report of the Department Committee on the high cost of working class dwellings* (1921) Cmd 1447.
[6] *Third interim report of the Committee on New Methods of House Construction*, 1925.
[7] A.W. Cleeve Barr (1964) Progress report on achievement in industrialised housing. In: *System Building 2*, Interbuild.

Chapter 2

[1] For over thirty years, the statistics were published annually in *Housing and Construction Statistics*. The new *Construction Statistics Annual* was first published in October 2000. It excludes the housing data of the original (now published separately) but includes details on government construction plans, PFI schemes and lottery projects.

Chapter 3

[1] *Construction Statistics Annual 2000*, Tables 3.1, 3.3, 3.4.
[2] *Building* 20 July 2001, pp. 39–53.
[3] Michael Ball (1988) *Rebuilding Construction – Economic Change in the British Construction Industry*, Routledge, London.
[4] Peter Mason, Chance of a lifetime *Building* 2 March 2001, p. 57.

5 Alfred McAlpine press release, 14 August 2001.
6 *Building* 7 May 1999, p. 24.
7 Patricia Hillebrandt (1971) *Small Firms in the Construction Industry*, HMSO, London.
8 R.R. Morton (1982) *The speculative housebuilding industry in the 1970s*, Unpublished PhD thesis, University of Liverpool.

Chapter 4

1 Cited in *Rethinking Construction*, para. 17.
2 Martin Briggs (1925) *A Short History of the Building Crafts*, Clarendon Press, Oxford, p. 25.
3 P.W. Kingsford (1973) *Builders and Building Workers*, Edward Arnold, London
4 See particularly, P.W. Kingsford *Builders and Building Workers*. The rest of this section depends heavily on Kingsford's text. One of the earliest histories was by R.W.O. Postgate *The Builder's History*, published in 1923 by the National Federation of Building Trade Unions.
5 Cited in *Builders and Building Workers*, pp. 106–107.
6 Cited in *Builders and Building Workers*, p. 178.
7 Quoted in W.S.J. Hilton (1968) *Industrial Relations in Construction*, Pergamon Press, London, pp. 200 foll.
8 Report of the Committee of Enquiry under Professor E.H. Phelps Brown into *Certain matters concerning labour in building and civil engineering*, HMSO, London, July 1968.
9 Construction Industry Training Board (2001) *Construction industry employment and training forecast 2001 – 2005.*
10 Westminster Business School: Education, Training and the Labour Market Research Group; *Standardisation and skills project* Draft summary report (full report due December 2001).
11 Health and Safety Executive *Health and safety statistics 1999/2000*, figs 1.3, 1.5, 1.7, 1.19.
12 e.g. Jennie Price 'Divided We Fall,' *Building* 18 August 2000, p. 25.
13 UCATT (2001) *Construction safety: building a new culture.*
14 See *Building*, 18 April 2000, p. 39.

Chapter 5

1 Spiro Kostof (1977) The practice of architecture in the ancient world in S. Kostof (ed.) *The Architect – Chapters in the History of a Profession*, Oxford University Press, New York.
2 F. Jenkins (1961) *Architect and Patron*, Oxford University Press, London, p. 55.
3 Mario Salvadori (1980) *Why Buildings Stand Up – the Strength of Architecture*, W.W. Norton, New York, p. 233.

[4] J.A. Picton (1875) *Memorials of Liverpool*, Longmans Green, Vol. 1, p. 564.

[5] F.M. Thompson (1968) *Chartered Surveyors – The Growth of a Profession*, Routledge & Keegan Paul, London, p. 182.

[6] D.S.E. Base (1982) Professional Engineers: the Integrating Individuals. In: *Future Needs – Civil Engineering Education*, ICE, p. 27.

[7] An argument in support of this view is elaborated in R. Morton and D. Jaggar (1995) *Design and the Economics of Building*, E. & F.N. Spon, London.

[8] Quoted in A. Satoh (1995) *Building in Britain*, Scolar Press, Aldershot, p. 97.

[9] Marion Bowley (1966) *The British building industry – four studies in response and resistance to change*, Cambridge University Press, Cambridge.

[10] J. Marston Fitch (1973) *American Building: The Historical Forces that Shaped It*, Shocken Books, New York, p. 126.

[11] Quoted in Jenkins *Architect and Patron*.

[12] Quoted in Jenkins *Architect and Patron*.

[13] C. Harding Down with Designers, *Building* 25 June 2001, p. 33.

[14] Reining in the Trójan Horses *Building* 16 March 2001, pp. 24–25.

Chapter 6

[1] Marian Bowley (1966) *The British Building Industry – Four Studies in Response and Resistance to Change*, Cambridge University Press, p. 352.

[2] C.W. Chalkin (1974) *The Provincial Towns of Georgian England – a Study of the Building Process 1740–1820*, Edward Arnold, London, p. 193.

[3] Quoted by John Harvey (1975) *Mediaeval Craftsmen*, B.T. Batsford, London.

[4] Linda Clarke (1992) *Building Capitalism – Historical Change and the Labour Process in the Production of the Built Environment*, Routledge, London.

[5] Quoted in Satoh *Building in Britain*, p. 291.

[6] Quoted in Satoh *Building in Britain*, p. 293.

[7] Quoted in Satoh *Building in Britain*, p. 43.

[8] Quoted in Satoh *Building in Britain*, p. 12.

[9] Quoted in Satoh *Building in Britain*, p. 63.

[10] Quoted in Satoh *Building in Britain*, p. 297.

[11] Quoted in Thompson *Chartered Surveyors – The Growth of a Profession*, p. 72.

[12] Quoted in Thompson *Chartered Surveyors – The Growth of a Profession*, p. 81.

[13] Quoted in Thompson *Chartered Surveyors – The Growth of a Profession*, p. 85.

[14] Quoted in Satoh *Building in Britain*, p. 92.

[15] J.A. Gotch (1934) *The Growth and Work of the Royal Institute of British Architects*, RIBA.

[16] P.W. Kingsford (1973) *Builders and Building Workers*, Edward Arnold, London.

Chapter 7

[1] Michael Ball (1988) *Rebuilding Construction – Economic Change in the British Construction Industry* Routledge, London.
[2] A description of CPI is contained in R.R. Morton and D. Jaggar (1995) *Design and the Economics of the Built Environment*, E. & F.N. Spon, London, pp. 301–305.
[3] The Comptroller and Auditor General (2001) *Modernising Construction* The Stationery Office, London.
[4] The Comptroller and Auditor General (2001) *Modernising Construction* The Stationery Office, London, Appendix 9.

Chapter 8

[1] Cited in *Rethinking Construction*, para. 92.
[2] Cited in *Rethinking Construction*, para. 31.
[3] Akira Satoh in *Building in Britain* gives a detailed and illustrated description of these developments in the nineteenth century.
[4] Cited in *Building in Britain* p. 235.
[5] These examples are quoted from R. Morton and D. Jaggar (1995) *Design and Economics of Building* E. & F.N. Spon which goes into greater detail.
[6] D.R. Harper (1990) *Building – the Process and the Product* CIOB 1978 (reprinted 1990).
[7] R.M.E. Diamant (1964) *Industrialised Building: 50 International Methods* Architect and Building News, p. 7.
[8] King cash *Building* 4 March 1994, p. 26.
[9] King cash *Building* 4 March 1994, p. 332.
[10] *Building Special Supplement* Canary Wharf 1991.
[11] John Bennett (1993) Managing Construction *Building Anniversary Issue*, p. 183.

Chapter 9

[1] The application of these techniques in energy saving methods is discussed in more detail in Morton and Jaggar (1995) *Design and the Economics of the Built Environment*, E. & F.N. Spon, London. The Building Research Establishment has published very many case-studies over the years illustrating their application in practice.
[2] Reported in *Building* 31 March 2000.
[3] Breath test *Building* 6 June 1997, pp. 43–50.
[4] Jolly green giant *Building* 23 March 2001, pp. 56–58.
[5] Greenpeace (1999) *Re-Source – Market Alternatives to Ancient Forest Destruction*, Greenpeace International Publications, Utrecht.
[6] G.H. Wytze, Vander Naald and Beverley G. Thorpe (1998) *PVC Plastic – A Looming Waste Crisis*, Greenpeace International Publications, Utrecht.

[7] Thomas Sharp (1940) *Town Planning*, Faber and Faber, London, pp. 15–16.

[8] Ebenezer Howard (1902) *Garden Cities of Tomorrow*, republished by Faber and Faber, London.

[9] Cost study *Building* 17 August 1999, pp. 90–95.

[10] The most energy efficient building ever *Building* 20 October 2000, p. 24.

[11] As green as it gets *Building* 4 February 2000, pp. 39–41.

Chapter 10

[1] C.C. Knowles and P.H. Pitt (1972) *The History of Building Regulation in London 1189–1972* H.E. Warne, London, p. 15.

[2] Death traps were council's fault, says quake expert, *Building* 3 September 1999, p. 13.

[3] Mowlem rides into cowboy country *Building* 10 July 1998, pp. 18–19.

[4] Public prefers cowboys *Building* 20 April 2001, p. 14.

Selected Bibliography

Books listed below have been chosen because they each expand on particular topics discussed or raised throughout this text. With the exception of some of the earlier historical studies and reports, they are all currently in print.

Adams, A. & Watkins, C. (2002) *Greenfields, Brownfields and Housing Development* Blackwell Science, Oxford.

Anderson, J., Shiers, D. & Sinclair, M. (2001) *The Green Guide to Specification 3rd edition* Blackwell Science, London.

Atkin, B., Borbrant, J. & Josephson, P. (2002) *Construction Process Improvement* Blackwell Science, Oxford

Aouad, G., Tah, J., Alshawi, A., Underwood, J. & Faraj, I. (2002) *Construction Project Information* Blackwell Science, Oxford.

Ball, Michael (1988) *Rebuilding Construction: Economic Change in the British Construction Industry* Routledge, London.

Banwell Report (1964) *The Placing and Management of Contracts for Building and Civil Engineering Work* Ministry of Public Building and Works, HMSO, London.

Bowley, Marian (1966) *The British Building Industry – Four Studies in Response and Resistance to Change* Cambridge University Press, Cambridge.

Briggs, M. (1925) *A Short History of the Building Crafts* Clarendon Press, Oxford.

Building (1993) *150th Anniversary Supplement* The Building Group, London.

Chalkin, C.W. (1974) *The Provincial Towns of Georgian England – A Study of the Building Process 1740–1820* Edward Arnold, London.

Chappell, D. (2001) *Standard Form of Building Contract 3rd Edition* Blackwell Science, Oxford.

Clarke, L. (1992) *Building Capitalism – Historical Change and the Labour Process in the Production of the Built Environment* (parts 2 & 3) Routledge, London.

Committee on the High Cost of Working Class Dwellings (1921) *Report* Cmd 1447.

Committee on New Methods of House Construction (1925) *Third Interim Report.*

Comptroller and Auditor General (2001) *Modernising Construction.*

Construction Industry Training Board (2001) *Construction Employment and Training Forecast 2001–2005* CITB.

Cox, S. & Clamp, H. (1999) *Which Contract? 2nd edition* RIBA Publications.

DeCamp, L.S. (1960 & 1990) *The Ancient Engineers* Dorset Press, New York.

Diamant, R.M.E. (1964) *Industrialised Building – 50 International Methods* Architect and Building News.

Doughty, M. (ed.) (1986) *Building the Industrial City* Leicester University Press, Leicester.

Edwards, B. & Turrent, D. (2000) *Sustainable Housing – Principles and Practice* E. & F.N. Spon, London.

Egan Report: Sir John (1998) *Rethinking Construction: The report of the Construction Task Force to the Deputy Prime Minister on the Scope for Improving the Quality and Efficiency of UK Construction* The Stationery Office, London.

Egglestone, B. (2000) *The New Engineering Contract* Blackwell Science, Oxford.

Emmerson Report (1962) *Survey of the Problems Before the Construction Industries* HMSO, London.

Garnham Wright, J.H. (1983) *Building Control by Legislation – the UK experience* John Wiley & Sons, London.

Gotch, J.A. (1934) *The Growth and Work of the Royal Institute of British Architects* RIBA.

Graham, P. (2002) *Building Ecology – Sustainability in the Built Environment* Blackwell Science, Oxford.

Graham, P. (2002) *Sustainable Construction in Action: 50 Practical Examples* Blackwell Science, Oxford.

Greenpeace (1999) *Re-Source: Market Alternatives to Ancient Forest Destruction* Greenpeace International Publications, Utrecht.

Greenpeace (1999) *Buying Destruction – A Report for Corporate Consumers of Forest Products* Greenpeace International Publications, Utrecht.

Greenpeace UK (1996) *Building the Future – a Guide to Building without PVC* Greenpeace UK, London.

Harper, D.R. (1978 & 1990) *Building – The Process and the Product* Chartered Institute of Building, Ascot.

Harper, R.H. (1985) *Victorian Building Regulations* Mansell Publishing, London.

Harvey, J.H. (1975) *Mediaeval Craftsmen* B.T. Batsford, London.

Hillebrandt, P. (1971) *Small firms in the Construction Industry* HMSO, London.

Hillebrandt, P. (1984) *Analysis of the British Construction Industry* Macmillan, Basingstoke.

Hillebrandt, P. & Cannon, J. (eds) (1990) *The Modern Construction Firm* Macmillan, London.

Hilton, W.S. (1968) *Industrial Relations in Construction* Pergamon, London.

Holgate, Alan (1986) *The Art in Structural Design – An Introduction and Sourcebook* Clarendon Press, Oxford.

Holt, A.S.-J. (2001) *Principles of Construction Safety* Blackwell Science, Oxford.

Jenkins, F. (1961) *Architect and Patron* Oxford University Press, London.

Kingsford, P.W. (1973) *Builders and Building Workers* Edward Arnold, London.

Knowles, C.C. and Pitt, P.H. (1972) *The History of Building Regulation in London 1189–1972* H.E. Warne, London.

Kostof, S. (ed.) (1977) *The Architect – Chapters in the History of a Profession* Oxford University Press, New York.

Langford, D. & Murray, M. (eds) (2002) *Construction Reports 1944–98* Blackwell Science, Oxford.

Latham Report: Sir Michael Latham (1994) *Constructing the team: final report of the government/industry review of procurement and contractual arrangements in the UK construction industry* HMSO, London.

Marston Fitch, J. (1973) *American Building: the Historical Forces that Shaped It* Shocken Books, New York.

Ministry of Works (1943) *Training for the Building Industry* Cmd 64828 HMSO, London.

Morton, R.R. & Jaggar, D. (1995) *Design and the Economics of Building* E. & F.N. Spon, London.

Pannell, J.P.M. (1964) *Man the Builder an Illustrated History of Engineering.* Thames & Hudson Ltd, London (reprinted Book Club Associates 1977)

Pawley, M. (1992) *Design Heroes – Buckminster Fuller* Grafton – Harper Collins, London.

Phelps Brown, E.H. (1968) *Report of the Committee of Enquiry under Professor F.H. Phelps Brown into Certain Matters Concerning Labour in Building and Civil Engineering* HMSO, London.

Powell, C.G. (1980) *An Economic History of the British Building Industry 1815–1979* Architectural Press Ltd, London.

Powell-Smith, V. & Billington, M.J. (1999) *The Building Regulations – Explained and Illustrated 11th edn*, Blackwell Science, Oxford.

Salvadori, M. (1980) *Why Buildings Stand Up – the Strength of Architecture* W.W. Norton & Co., New York.

Satoh, A. (1995) *Building in Britain – The origins of a modern industry* Scolar Press, Aldershot.

Simon Report (1944) on *The Placing and Management of Building Contracts*, Ministry of Works HMSO, London.

Stroud, D. (1966) *Henry Holland: His Life and Architecture* Country Life, London.

Summerson, J. (1973) *The London Building World of the Eighteen Sixties* Thames & Hudson, London.

Summerson, J. (1980) *The Life and Work of John Nash, Architect* MIT Press, Cambridge, MA.

Sunley, J. & Bedding, B. (eds) (1985) *Timber in Construction* Batsford/TRADA, London.

Thompson, F.M.L. (1968) *Chartered Surveyors – The Growth of a Profession* Routledge & Keegan Paul, London.

University of Westminster (2001) *Innovation and skills: a transnational study of skills, education and training for prefabrication in housing* Unpublished report for EPSRC and DETR, summarised in *Building* 'Learning Difficulties', 8 June 2001, pp. 38–41.

UCATT (2001) *Construction Safety – Building a New Culture* Union of Construction, Allied Trades & Technicians, London.

Vale, B. & Vale, R. (1991) *Green Architecture: Design for a Sustainable Future* Thames & Hudson, London.

Index